The Skull Collectors

Race, Science, and America's Unburied Dead

The Skull COLLECTORS

ANN FABIAN

The University of Chicago Press :: Chicago and London

The University of Chicago Press, Chicago 60637
The University of Chicago Press, Ltd., London

Published 2010
Paperback edition 2020
Printed in the United States of America

29 28 27 26 25 24 23 22 21 20 1 2 3 4 5

ISBN-13: 978-0-226-23348-2 (cloth)
ISBN-13: 978-0-226-76057-5 (paper)
ISBN-13: 978-0-226-23349-9 (ebook)
DOI: https://doi.org/10.7208/chicago/9780226233499.001.0001

LIBRARY OF CONGRESS CATALOGING-IN-PUBLICATION DATA

Fabian, Ann.
The skull collectors: race, science, and America's unburied dead / Ann Fabian.
p. cm.
Includes bibliographical references and index.
ISBN-13: 978-0-226-23348-2 (cloth: alk. paper)
ISBN-10: 0-226-23348-0 (cloth: alk. paper) 1. Morton, Samuel George, 1799–1851. Crania Americana. 2. Craniology—Social aspects—United States— 19th century. 3. Anthropometry—United States—History—19th century. 4. Skull—Catalogs and collections—United States—History—19th century.
I. Title.
GN90.F33 2010 200904771
599.9 48—dc22

∞ This paper meets the requirements of ANSI/NISO Z39.48-1992 (Permanence of Paper).

For Chris and Andrew and Isabelle

Contents

Acknowledgments

My thanks to colleagues and friends who have talked to me about this book and read portions of it: Elizabeth Kaspar Aldrich, Mia Bay, Herman Bennett, James Brooks, Joshua Brown, Scott Casper, Kathy Compagnon, Jay Cook, Catherine Corman, Leah Dilworth, Adam Domby, Brad Evans, John Mack Faragher, Richard W. Fox, Ed Gray, Karen Halttunen, Alison Isenberg, David Jaffee, Toby Jones, Joel Kuipers, T. J. Jackson Lears, Jill Lepore, Jan Lewis, Julie Livingston, Fran Mascia-Lees, Matt Matsuda, Meredith McGill, Susan McKinnon, Teresa Murphy, Carla Nappi, Bill Reese, David Shields, Ben Sifuentes-Jauregui, Mel Stein, Keith Wailoo, Michael Warner, Carla Yanni, and especially Marni Sandweiss, steadfast intellectual companion. I hope my long-ago research assistant Kate Sell is happy to see that she did not waste those weeks she spent organizing files and transcribing letters.

A fellowship from the John Simon Guggenheim Memorial Foundation gave me a chance to begin writing this book; a Rutgers sabbatical, the time to finish it. A fellowship from William Y. and Nettie K. Adams let me spend summer weeks at the School for Advanced Research exploring the history of physical anthropology. I am grateful to Anne Palkovich for telling me what to read. Several years ago, Janet Monge showed me the remains of Morton's collection at the Museum of Archaeology and Anthropology at the University of Pennsylvania.

I would not have been able to retrieve this book's stories from the archives without help from librarians at the Academy of Natural Sciences of Philadelphia, the American Museum of Natural History, the American Philosophical Society, the Library Company of Philadelphia, the National Anthropological Archives of the Smithsonian Institution, the New-York Historical Society, the New York Public Library, and the Peabody Essex Museum. I am especially grateful to Arlene Shaner at the New York Academy of Medicine, to George Miles at the Beinecke Rare Book and Manuscript Library, and to Jaclyn Penny and Georgia Barnhill at the American Antiquarian Society, who found illustrations for me.

Audiences at the College of Charleston, the Huntington Library, the Mailman School of Public Health at Columbia University, Michigan State University, the Department of American Studies at Rutgers-Newark, the University of Delaware, the University of Michigan, the University of Nevada at Reno, the University of Pennsylvania, the University of Southern California, and Yale University listened to early versions of my research and nudged it in better directions. A gift to Rutgers University from Mr. and Mrs. John J. Byrne gave me the opportunity to teach a Byrne Family First-Year Seminar on "Curiosity," which provided funds to pay for the book's illustrations. I am also grateful to Rutgers for supporting faculty seminars in the humanities and pleased to acknowledge the conversations at the Center for Cultural Analysis on "Secularism" and "Chance and Luck" that helped shape this book. For two years Jackson Lears and I led a seminar on "The Question of the West" at the Rutgers Center for Historical Analysis. Our Tuesday morning discussions reminded me each week that I had an intellectual life despite the administrative tasks I had taken on in the dean's office. For that, I am deeply grateful to Jackson and to all the seminar's participants.

Thanks also to Wendy Strothman, who took me into her stable of writers; to Robert Devens, who championed the book through the University of Chicago Press; to Michael Sappol and Ben Reiss, who challenged me with very smart comments on the manuscript; and to Erin DeWitt, who corrected it so well. And finally, thanks to my family—Isabelle Smeall, Andrew Smeall, and Chris Smeall—who read paragraphs, transcribed letters, took pictures, trooped along to visit Columbus's graves in Seville and Santo Domingo, and gave me many things to think about besides skulls. I also want to give a nod to an anonymous woman from Chicago who left an old joke on my office answering machine. An American tourist was rummaging through an antique shop in Ireland and found a skull. "Whose skull is this?" he asked the shop owner. "Why, that's the skull of the bandit Paddy McGinty." "I'll buy it," the tourist said, and went back to rummaging. He found another skull, much smaller. "And whose skull is this?" "Why that's the skull of Paddy McGinty when he was a little boy." It was good to get a joke. I saved it as long as I could.

For as every one knows, ghosts of the unburied dead haunt the earth and make themselves exceedingly disagreeable, especially to their undutiful relatives.

—SIR JAMES FRAZIER, *On Certain Burial Customs as Illustrative of the Primitive Theory of the Soul* (1885)

INTRODUCTION

Ghosts of the Unburied Dead

Naturalist Samuel George Morton collected skulls: animal skulls and human skulls. He had a mouse, a finch, a hippopotamus, and a Seminole man drilled by a bullet in his left temple. By the time Morton died in 1851, he had nearly a thousand human crania, a collection many thought was the largest collection in the world. He complained that he had paid out a small fortune in shipping charges for the skulls that friends and acquaintances had been sending for nearly twenty years. But the collection, with its hundreds of Native American skulls, made him famous in the naturalist community. Friends teased him, calling his Philadelphia study an "American Golgotha."

Morton earned a reputation as a steady hardworking man. He sorted his skulls into racial groups and then measured them. Morton's skulls launched American work in craniology and mapped out the contours of a distinctive American inquiry that involved thinking about race, particularly the racial characteristics of Africans, and collecting dead bodies, particularly the bodies of Native Americans. He subscribed to the widely held belief that there were five races—the Caucasian, the Mongolian, the American, the Malay, and the Ethiopian—but then concluded that each race represented a different species created for one of the earth's continents, an idea that set him at odds with clergymen and believers who were certain that all men were the children of Adam.

Some found it easier to accept another of his ideas. While anatomists and naturalists had studied variation in the shape of skulls and sketched their facial angles, Morton measured their volume and worked out an average cranial capacity for each of the races. By his calculations, Caucasian skulls were the largest. His next step proved more consequential. The larger the cranial capacity, the larger the brain, Morton thought. And the larger the brain, the better the man. Over the years, many have challenged Morton's methods and his conclusions, but his numbers gave him confidence to conclude that the Caucasian "race is distinguished for the facility with which it attains the highest intellectual endowments." His findings jibed nicely with the popular prejudices of many of his contemporaries and appeared to give empirical foundation to loose speculations about racial origins and racial differences.[1]

Morton's work on skulls secured his place in the annals of what historians have called "scientific racism." In fact, Morton helped define the term, exploring the differences among men with the detached neutrality we associate with scientific inquiry. A friend described him at work. Each skull underwent an "ordeal of examination," as Morton made "accurate and repeated measurements of every part." "Where a case admitted of doubt, I have known him to keep the skull in his office for weeks, and, taking it down at every leisure moment, sit before it, and contemplate it fixedly in every position, noting every prominence and depression, estimating the extent and depth of every muscular or ligamentous attachment, until he could, as it were, build up the soft parts upon their bony substratum, and see the individual as in life."[2]

It's a compelling image, inspiration for a portrait painter who might include Morton in a long line of men contemplating their certain bony futures. But Morton's ideas and Morton's skulls turned into troubling legacies at the end of the twentieth century. Scholars studying genetics unearthed his work as they returned to debates on the biological bases of racial differences.[3] His skulls, which for decades had been moldering away in storage at Philadelphia's Academy of Natural Sciences, became objects of controversy too, as the descendants of those whose remains had been collected came, took them home, and buried them. It is never easy for museums to give up the things they hold, but attitudes about the collection and display of human remains shifted over the course of the twentieth century. A collection like Morton's appeared to be a residue of colonialism and conquest, the by-product of an outmoded science.[4]

These controversies brought me to Morton and his collection, and through the man and his skulls to the lively, contradictory cultural world of the United States in the second quarter of the nineteenth century. This was a world where Morton could launch an ambitious scientific project, publish a book with beautiful lithographs of his specimens, worry about money, toy with phrenology, correspond with members of learned societies in Europe, and send his collectors to gallows and battlefields where men had died or to the mounds, caves, and burial grounds where Native Americans had buried their dead. We know something about that world through the eyes of a young Walt Whitman, through the writings of Emerson, Thoreau, Poe, and Melville, all men who shared something of the impulse to get their country down on paper.

But to begin with these encounters over dead bodies gives us a new point of entry into mid-century America, weaving together histories and stories that have been told separately. To recover traces of the experiences of those whose remains were collected and put on display, we need to think about Philadelphia naturalists and Methodist missionaries; diplomats, doctors, explorers, and the showman P. T. Barnum; Java, Oregon, Fiji, and Morton's Philadelphia; popular interest in Egypt, Frederick Douglass's attack on Morton's work, and the consequences of Indian Removal. The stories that begin with death and burial give us some new ways of investigating the disparities and inequalities that dogged individuals in their lives and followed them into graves or onto collectors' shelves.[5] The tensions between skulls as measures of racial difference and as markers of common humanity lie at the heart of this book.

This history unfolds in five chapters. A first chapter looks at Morton's life and investigates his skull collection. I was surprised by how much Morton and his collectors wrote about their work with human remains. Their correspondence is full of chatter, as though they needed to acknowledge they were doing something that set them apart from superstitious friends and neighbors who were still spooked by dead bodies.[6] Grave robbing gave him a "rascally pleasure," one man told Morton. And it was risky too, particularly when mourners chased collectors away from corpses. Still, Morton's collectors persisted, keeping up a steady supply of heads from America and around the world.

In some cases, circumstances made heads hard to get, as a story about a young Chinook man from Oregon makes clear. This man, Stumanu, met Morton in Philadelphia late in the winter of 1838. The long string of events that brought these two men together takes up the second chapter. The fact that Stumanu had a remarkably flattened head made him a marvelous spectacle for skull-measuring craniologists and skull-reading phrenologists. They found skulls like his fascinating and paid a premium for skulls that had been intentionally flattened in infancy by cradle boards or bandages. Morton described his meeting with Stumanu in his *Crania Americana; or, A comparative view of the skulls of various aboriginal nations of North and South America*, the sumptuous illustrated volume he published in 1839. *Crania Americana* is Morton's most important publication, a striking example of American intellectual enterprise and Philadelphia printing and lithography. The third chapter describes the history of the book's production and circulation. *Crania Americana* is a contradictory volume, combining the dry empiricism of Morton's descriptions of racial difference with richly textured lithographs of skulls that he commissioned from a talented local artist.

The book's illustrations remind us that although Morton aspired to strip collected skulls of symbolic meanings, his work could not proceed in isolation either from the old visual language of skulls that reminded viewers of their mortality or from the racial violence of 1830s and 1840s America, both in the slave states and in Morton's Philadelphia. In 1831, about the time Morton began to collect skulls, Nat Turner led fellow slaves in a rebellion against slave owners in southern Virginia. Panicked slaveholders executed suspected plotters and stuck their heads on stakes, a warning to those who would join the rebels. There is no direct connection between these staked heads and Morton's collection, but on occasion the violence and humiliation of such displays reached his quiet study. In the 1840s, a herpetologist working in Liberia sent Morton heads of African tribal leaders who led a bloody resistance to settlement on their lands by former American slaves. He had removed the rotting heads from the stakes where they had been posted to frighten others and had sent them to Philadelphia, where Morton cataloged them as specimens of their tribes.

That process of turning a man into a specimen is the specific focus of the fourth chapter. Skulls of Pacific Islanders were especially rare in American collections. The United States Exploring Expedition brought a man from Fiji to New York in 1842, and an expedition naturalist thought Morton would be interested in his skull. The man lived only a few hours in New York. Hospital workers buried his body in Brooklyn but sent his head to Washington, where it could be displayed along with the expedition's extraordinary collections. Nothing in his experience growing up Fiji could have helped him picture a death so far from home and such a bizarre disposition of his corpse. His story is unusual, but collections of human remains are filled with bodies of men and women who, like him, died far from the communities they might have expected to protect their corpses, to usher them out of one life and into a next.[7]

Unprotected corpses began to haunt Morton's old friends and neighbors when the grim toll of America's Civil War overwhelmed the country's customary ways of handling the dead. Morton had been a decade dead when the country went to war over slavery, an institution he never defended outright, even though his work left traces in the intellectual architecture of slavery's defenses. The war's death toll changed the ways that Morton's contemporaries handled dead bodies. Grieving men and women eased their losses by imagining that friends, sons, brothers, fathers, lovers, and husbands had died in sacrifice to a greater cause. That sacrifice left a country awash in bodies.[8] The war was a boon for embalmers and undertakers, who preserved the remains of dead soldiers and shipped them home to grieving families. It was also a boon for craniologists, who finally had a good supply of "white" men's skulls.

But the wartime impulse to measure bodies and the postwar push to sort corpses and number the dead had very different consequences for Native Americans, particularly those on the western plains. While government officials sorted the war dead and settled their bodies into cemeteries in the East, United States army surgeons, posted in the West, unburied the remains of hundreds of Native Americans and sent skulls, skeletons, and body parts to museums in Washington. One rarely imagines that the somber work of reburying the Civil War dead and the scientific work of unburying and collecting Native American remains ran on parallel tracks, but the two drew on the same personnel and employed the same ideas and the same tools, albeit to very different ends for communities of survivors. Native American remains collected in the years immediately after the Civil War set the stage for late twentieth-century efforts to rebury the dead.[9]

Those efforts helped uncover the cultural history that has gone into this book. I tried to avoid the gallows humor that runs through a lot of the writing on dead bodies, although it is sometimes difficult to steer clear of the rascally eccentricities of passionate skull collectors.[10] That dark humor sometimes masks dimensions in the collection and display of large numbers of human dead—not for their kin or descendants and not in a manner the living could ever have imagined. Morton and his collectors picked up the remains of communities devastated by war, disease, or poverty, and worked most easily where people had been unable to bury their dead or unable to watch over corpses.

As Morton's collectors unburied the dead, they left a landscape riddled with emptied graves. Those empty graves draw us back into the communities of those whose remains filled collectors' shelves. I have traced a few of those stories, but I have worked on them long enough to know that empty graves and the unburied dead trouble survivors. Melville shows this sharp grief in an early chapter of *Moby-Dick*. Ishmael visits the Whaleman's Chapel in New Bedford, Massachusetts, where silent men and women stare at walls lined with marble tablets "sacred to the memory" of men lost at sea. "Oh! Ye whose dead lie buried beneath the green grass; who standing among flowers can say—here, *here* lies my beloved; ye know not the desolation that broods in bosoms like these. What bitter blanks in those black-bordered marbles which cover no ashes!" The passage is heavy-handed, but it captures the special grief in stories of the unburied dead. I hope that a collection of those stories will begin to put words into the bitter blanks that skull collectors left on emptied graves.

42. CELTIC Irishman, aged 21, imprisoned for larceny, and in all respects a vicious and refractory character. Died A.D. 1831. I.C. 97. [Approaches the square Germanic form.]

59. ANGLO-SAXON head; skull of Pierce, a convict and cannibal, who was executed in New South Wales, A.D. 18__. F.A. 85°. I.C. 99. [A long and strictly oval head.]

427. CHINESE, hanged for forgery at Batavia, in Java: man, ætat. 30. F.A. 78°. I.C. 83. Dr. Doornik.

—J. AITKEN MEIGS, *Catalogue of Human Crania, in the Collection of the Academy of Natural Sciences of Philadelphia* (1857)

CHAPTER 1

"The Promise of a Fine Skull"

"Death of Dr. Morton"

Samuel George Morton died on Thursday, May 15, 1851. He was fifty-two. The stoop-shouldered skull collector had never been a robust man, but no one had expected him to die that spring. He went to bed on Saturday night, complaining that his head hurt. On Sunday morning, his legs were numb and pain shot down his back. He slept through the night on Tuesday and began talking again on Wednesday morning. The family hoped the good Tuesday rest had rallied him. But he died Thursday morning.

Family and Philadelphia friends would miss Morton. He was an important figure in that city's public life, helping to keep up Benjamin Franklin's intellectual legacy in the half century after Franklin's death and helping to promote American science among European intellectuals. Morton and his friends had eclectic interests, amateurs' tastes for a variety of subjects. He and his fellow naturalists studied fossils, frogs, birds, rats, the age of the earth, and the nature of electricity. They were curious about people like a young Chinook man with his flattened head who came to town with a Methodist missionary in 1838. Morton could measure a man one day, write about a hippopotamus the next, and then sit down to sketch a hermaphrodite orangutan. (The day he took sick, he had gone to a lecture on mummies.)

To later generations of university-trained scientists, Morton and his friends seemed eclectic dabblers, although a generous curiosity shaped Morton's intellectual world. And this openness suited him well. Despite poor health and an occasional bad mood, he led the full life of a learned middle-class Philadelphian. There were patients to see, letters to answer, papers to read, lectures to attend, and people to meet at the Academy of Natural Sciences, the Almshouse Hospital, and the American Philosophical Society. On Sundays, Morton was "at home," welcoming friends and acquaintances for evenings of conversation. And without having to leave the city, he corresponded with collectors and fact gatherers, scattered around the country and around the world.[1]

Work with skulls gave Morton a special standing, even in his close group of naturalist friends. Many contemporaries thought Morton's collection of human skulls was strange and somewhat creepy, but its very strangeness gave Morton's work an extra dose of prestige. He was disciplined, objective, and empirical, cultivating a calm stance of a detached observer, as he analyzed objects that made others squeamish. There would have been less harm in this intellectual work if in the decade after Morton's death slavery's defenders had not turned his observations from descriptions of differences in the shape, size, and form of skulls in humans and animals into social commentary on the differences among men. Morton's training as a comparative anatomist taught him to discern small differences, distinctions that lurked beneath the surface or skin of things. But these questions about differences among skulls pitched Morton headlong into debates on the origin and nature of races. Morton's legacies have troubled his place in the history of American science. He left his wife his collection of one thousand human skulls, and some of his associates used his skull work to construct the peculiar intellectual legacy known as "scientific racism."[2]

Morton's family and friends weren't thinking about science or race that May morning. They were grieving for a husband, father, and colleague. By standards of the day, Morton's was a good death; he died at home, with his family, and after a brief illness. While family members imagined the state of his soul, naturalist colleagues wondered about his body. What disease could have killed their friend so quickly? Could they find its signs in his corpse? Rebecca Morton, the longtime wife of a naturalist, must have been accustomed to just such questions, and with her permission, six of Morton's doctor friends slid the corpse from the deathbed onto a dissection table and set to work on an autopsy.

They probably waited until Friday and started in the morning when the slanting spring sun gave the best light to see the inner organs of their friend's fresh body. They began with the head. Peeling back the scalp and cutting into the skull, they found the blood clots and burst blood vessels that had set off the "*meningeal apoplexy*" that had killed Morton. Despite the damaging stroke, friends found that the "brain itself was large and symmetrical"—a comment sure to please a man who had spent much of his career laying the foundations for the argument that superior men had large skulls and large brains. They worked their way down the body, slicing through Morton's chest and pulling out his "large and flabby" heart. His liver looked healthy, but his spleen was "soft and pulpy," and a bout with pneumonia had shriveled the left lung.[3] Work done, they stitched up Morton's body and gave it to Mrs. Morton for "laying out." Perhaps she shaved his face and dressed him in a good suit, performing those widow's tasks with her daughters, sisters, or neighbors. (A generation later, professional undertakers took on this work.)

There are interesting features in the story of Morton's dissection, none as striking to our ironic sensibilities as the fact that no one took Morton's head, boiled off its flesh, and added it to his collection. The collector's skull on a shelf would have been a nice way to finish off his great work on crania, a signature, of sorts—the collector collected. While it seemed right for Morton's friends to dissect his body, they would never have taken his head, at least not in 1851 and not for this collection. Some of Morton's contemporaries knew that Oliver Cromwell's skull was at Oxford's Ashmolean Museum of Art and Archaeology and that Descartes' skull, making its way through European collections, had sold in 1820 as part of a Stockholm library. Travelers had seen saints' skulls on church visits in Italy and France.[4] These were celebrated heads—heads of geniuses, martyrs, and holy men and women.

Naturalists in Protestant America, however, operated by a different logic. Morton and his colleagues bragged that they had outgrown attitudes about dead bodies that afflicted superstitious contemporaries. Still, they did not apply those new attitudes to the corpses of families and friends. The living Morton had been a thinking subject, and he did not leave behind a body as an object to be collected or displayed. There were clear distinctions between Morton the collector and those whose remains he collected. These distinctions carry a dark history of past practices into even the most scientific collections of human remains, a reminder that to collect and display the dead humiliated defeated enemies.

A trace of ancient practices appears in Morton's modern world. A British social psychologist explored a spike in the traffic in body parts in the middle years of the nineteenth century. The great enthusiasm for scientific collecting that coincided with European wars on imperial frontiers in Africa, India, Australia, and the Americas suggests that a case can be made that the work of learned men encouraged white soldiers to give in to a temptation to collect parts from the bodies of those they had killed. Scientific collecting provided an alibi for some who might once have paused before taking a head, wondering if they had fallen from the grace of "civilized warfare" and into savage and sacrilegious ways. The special urging of a man like Morton on his search for specimens covered that "primitive urge" to take trophies. In the 1850s, there was no room for the head of a good man like Morton on his shelves.[5]

Philadelphia had a better place for Morton's corpse. In the spring of 1851, family and friends accompanied Morton's body three miles up the Schuylkill River to Laurel Hill Cemetery, the "Garden Cemetery" that prosperous Philadelphians had constructed in the mid-1830s. Prominent citizens, including the director of the Library Company of Philadelphia and members of the Academy of Natural Sciences, designed Laurel Hill to appeal to middle-class Philadelphians, like the Mortons. The jumble of headstones cluttering old churchyard burial grounds exuded the romance of ruins, but these messy graveyards were not places to deposit the modern dead.

New "rural cemeteries" offered something better. Travelers returned from Europe with good reports of Paris's Père-Lachaise cemetery, and Americans began to build a new kind of burying ground. The Massachusetts Horticultural Society built the first of the new garden cemeteries, Mount Auburn Cemetery in Cambridge. Over the next three decades, other cities followed: Greenmount in Baltimore, Green-Wood in Brooklyn, Spring Grove in Cincinnati, Allegheny in Pittsburgh, Bellefontaine in St. Louis, Oakland in Atlanta, Calvary in Milwaukee, and Graceland in Chicago. Neat graves hid the rot that city dwellers associated with

Fig. 1. Samuel George Morton's monument
(Photo courtesy of Carol Yaster, Laurel Hill Cemetery.)

older cemeteries, and the garden cemeteries smelled of grass and flowers. Even now, with their great trees and generous views, they seem gifts from landscaping geniuses of another century.[6]

Laurel Hill was Philadelphia's garden cemetery. It expressed a sense of order meant to please wealthy white Philadelphians. Restrictions kept African Americans from buying plots and their white employers and white relatives from burying black bodies at Laurel Hill. Cemetery directors encouraged families to erect grave markers carved in long-lasting granite. Mourners passed neighbors who had come to stroll the grounds, walking Laurel Hill's winding paths, appreciating budding trees and singing birds that robbed death of "half its horrors." May's violets were in bloom for Morton.[7]

Yet Morton's work skews this history too. Some among his mourners pictured his thousand skulls (his cabinet filled with unburied dead) while they lowered his corpse into a grave. Morton's family marked his grave with an obelisk, adopting a neoclassical design to express the secular piety of Morton and his friends, religious sentiments that mixed easily with prosperity, art, and science. Four faces of Morton's tombstone hailed: "Samuel George Morton (1799–1851)," "Physician, Naturalist, and Ethnologist," "President of the Academy of Natural Sciences," "Author of Crania Americana, etc., etc." (The literary output taxed the stonecutter, but had he continued he could have added Morton's publications on fossils, anatomy, pathology, and pulmonary consumption.) A last panel concluded: "Wherever Truth is Loved or Science Honored, His Name will be Revered."

Laurel Hill's winding paths lead visitors to a grave that has carried Morton's memory from the middle years of the nineteenth century to the first decades of the twenty-first. There is no move to unbury his body and put his skull on display. But it is worth pausing to appreciate the contrast between his well-buried remains and the unburied dead that filled his shelves.

Morton started collecting skulls seriously in 1830, when he was planning lectures on human anatomy for Philadelphia medical students. He set out to demonstrate "the difference in the form of the skull as seen in the five great races of men," he wrote. He followed the German anatomist Johann Friedrich Blumenbach, who sorted human races into the Caucasian, the Mongolian, the American, the Malay, and the Ethiopian. Morton's plan ran into a snag when he tried to find Mongolian and Malay skulls in Philadelphia. "Caucasian and Negro crania were readily procured, and two or three Indian skulls were placed at my disposal; but for the Mongolian and Malay I inquired in vain. I resolved, therefore, to supply this remarkable deficiency in an important branch of science; and much time, toil, and expense have been rewarded by the acquisition of 867 human skulls and 601 inferior animals," he wrote in a tally in 1849.[8]

He added a few dozen more skulls before he died. He left hundreds of Native American skulls, some odd-shaped heads from Sweden, skulls of a Finnish burglar, a German dwarf, an Australian cannibal, a Hindu "fanatic from Juggernaut," and a syphilitic man who died in a Calcutta hospital. An acquaintance sent him a Celtic skull he dug up on the battlefield at Waterloo; another sent him the skull of an Afghan boy killed in a massacre at Jugdalluk; others sent skulls from catacombs in Egypt and France, cemeteries in Peru, and churchyards in England and Ireland. Morton bought some skulls he labeled the heads of "filles de joie," and a doctor in Cuba sent him the skulls of fifty-five Africans—including girls and boys as young as twelve—who had died as slaves on a sugar plantation outside Havana. All in all, an extraordinary collection of diverse heads from around the world.

Morton wanted to calculate the average sizes of skulls of races, tribes, or groups, and his great innovation was to explore means to measure the internal capacity of skulls. Predecessors had sorted skulls by profile (the "facial angle" devised by Dutch artist anatomist Peter Camper), by length (the "norma verticalis" used by Blumenbach), or by views from the base or front. Morton measured empty skulls. But comparison was a dodgy business when he had little control over the samples he received. He took the skulls that came his way—mixing children and adults, men and women. He cleaned each one, coated it with varnish, measured its various angles, and calculated a volume by filling it with either liquid mercury (a toxic substance blamed for the madness of hatters but which also could explain the sulky moods that sometimes frightened Morton's friends and collaborators) or with pepper seed and buckshot.

We can picture Morton's work as messy and uncertain, but his success in reducing skulls to numbers is as significant as any of the conclusions he drew from those numbers. Although historians have condemned his contribution to "scientific racism," we can credit him with a contribution to modern methodology. He turned an unwieldy collection of skulls into information he could exchange with collectors, naturalists, and scholars. He recruited allies—the artists, phrenologists, printers, naturalists, and explorers who brought skulls to Philadelphia and helped move ideas about those skulls well beyond Morton's Philadelphia study. This "network" helped Morton establish a "natural history" of race that pretended to have "discovered" the racial differences that it, in fact, had helped to invent.[9]

Diplomats, travelers, soldiers, and doctors also helped fill his shelves with human and animal crania. Morton was lucky to sit at the tail end of the long Enlightenment project to catalog, classify, sort, and order the natural world, although he was among the first naturalists to make systematic collections of human skulls. He called on amateur naturalists who had been corresponding with Philadelphia naturalists since early in the century and took advantage of the old flow of ideas and specimens that had been running back and forth between Europe and the Americas for nearly two centuries. To many naturalists, America seemed rich enough in nature's resources to make up for its relatively poor culture.[10]

It should be clear from this short account that something more than curiosity about anatomy fueled Morton's collecting. He responded to questions from a field his contemporaries called "ethnology," a science that came into vogue in the 1840s to describe the work of those who compared "the characters of the different races of men." With the discipline of a trained naturalist, Morton described individual skulls, searching for skulls that he could class as typical of a race or national group. It was easy to be distracted by deformed skulls or skulls scarred by syphilis, bullets, axes, and surgeons' drills, but with a trained eye, Morton moved past these distractions to find the features that characterized a group. Phrenology's map of the brain gave him some ideas about where to look for "remarkable developments" in individual skulls, but phrenology with its conviction that better-developed organs enlarged areas of the skull led Morton to believe that the "internal capacity [of the skull was] indicative of the size of the brain" and then to argue that superior races had bigger skulls.[11]

American naturalists, who had read Thomas Jefferson's *Notes on the State of Virginia*, were primed to accept arguments about size. Jefferson wrote his *Notes* to counter the Count de Buffon's slight that "all living nature has become smaller" on the American continent. The continent's climate had produced feeble "savages" and reduced domestic animals. Others argued that Europeans who came to America degenerated too. Instead of dismissing Buffon's assumption that smaller creatures were inferior, Jefferson gathered evidence on the size and weight of America's quadrupeds. The argument sent American naturalists looking for gigantic things, like mastodons, to prove that the New World environment did not make deficient creatures.[12]

But that discussion of size had different consequences when it was applied to human skulls. Morton and his colleagues had a stake in this game; they wanted Caucasian heads like theirs to be the largest. Along with a passion for measuring, Jefferson's *Notes* also helped inspire the scientific study of racial differences. As Jefferson speculated, one could hazard the opinion that people of African descent "are inferior in the faculties of reason and imagination," but "to justify a general conclusion, requires many observations, even where the subject may be submitted to the Anatomical knife, to Optical glasses, to analysis by fire, or by solvents. How much more then where it is a faculty, not a substance, we are examining; where it eludes the research of all the senses; where the conditions of its existence are various and variously combined; where the effects of those which are present or absent bid defiance to calculation; let me add too, as a circumstance of great tenderness, where our conclusion would degrade a whole race of men from the rank in the scale of beings which our Creator may perhaps have given them."[13]

Here was a mystery to challenge the best naturalists. It is easy now to see social contours in Morton's project, to follow lines of race and class that made it possible to collect some skulls and not others. Skulls, collected as objects, came disproportionately from communities of Native Americans, from Africans and their descendents, from criminals, and from poor and displaced individuals who had lost ties to friends and relatives who would have cared for their corpses. In its social dimension, Morton's collecting practice resembled the dissection work of his anatomical colleagues, whose subjects came largely from poor communities—black, Irish, and immigrant.[14] And it is easy to dismiss Morton's work as outmoded "race science," but that label risks missing the more interesting intellectual and cultural currents that lit up his study. He called himself a "craniologist"—a man who understood the meaning of skulls. He called his work "craniometry"—the special branch of body measuring ("anthropometry") that concentrated on skulls; or "cranioscopy"—the visual study of skulls; or "craniography"—the written description of skulls.

"Physician, Naturalist . . ."

Samuel George Morton was born in Philadelphia in January 1799, the youngest of nine children. Only three survived past infancy. His father, an Irish Protestant merchant, died in the summer of 1800, leaving a widow with three young children. Mrs. Morton moved her family to a Quaker community in Westchester County, north of New York City, where she joined the Society of Friends. When the children were old enough, she enrolled them in Quaker schools.

The family moved back to Philadelphia in 1812 when Morton's mother married a nature-loving merchant whose special interests were geology and mineralogy. This man launched young Morton into Philadelphia's particular mix of nature and commerce. Morton remembered that he liked studying nature, although he complained that his stepfather and his Quaker schoolmasters sacrificed his poetic side to their practical commercial interests. (A shame, according to an enthusiastic eulogist, since Morton's poetic talent was drawn from "the same supersensitiveness we have seen in John Keats.") The stepfather apprenticed the boy to a merchant friend. Morton remembered hating his days hunched over columns of numbers and recalled stealing time to read history, natural science, and, most memorably, Benjamin Rush's *Sixteen Introductory Lectures to Courses of Lectures upon the Institutes and*

Fig. 2. Samuel George Morton (1799–1851)
(Emmet Collection, Miriam and Ira D. Wallace Division of Art, Prints and Photographs, The New York Public Library, Astor, Lenox and Tilden Foundations.)

Practice of Medicine, a summary of Dr. Rush's course at the University of Pennsylvania. Medicine led Morton back to natural history, but the merchant's world may have left a trace on his skull work, rubbing off as a taste for measurement, tabular records, and confidence that numbers offered an accurate accounting of human difference. And a merchant's skill with inventory helped collector Morton keep track of his skulls.[15]

Morton capitalized on his reading and trotted out what he had learned from Rush to impress the doctors who were taking care of his ailing mother in the late 1810s. One of them, Dr. Joseph Parrish, who ran a private medical school, saw a spark of scientific talent in the boy and invited Morton to attend his lectures on anatomy. Parrish's lectures drew crowds of bright young Philadelphians, including Richard Harlan, a young man who mentored Morton early in his career.

Harlan was an intellectual prodigy, producing a study of fossil vertebrates in 1815 that was good enough to prompt Philadelphia's senior naturalists to elect the nineteen-year-old to their Academy of Natural Sciences. Harlan's energy seemed a match for America's ambitions in those years following the War of 1812, and in a brief career he wrote on paleontology, neuroanatomy, zoology, and herpetology. In 1816, while still a medical student, he shipped as a surgeon to Calcutta, where he marveled at Indian religious devotions but expressed his Christian horror at a widow burning. He published a study of the human brain; described snakes, rats, salamanders, whales, and a manatee; and put his name on the *Basilosaurus Harlan*, a whale fossil he mistook for an extinct lizard.

When Morton arrived at Parrish's classroom about 1817, Harlan, just three years his senior, was leading Parrish's students through their anatomy lessons and lecturing after hours to paying audiences at Charles Willson Peale's Philadelphia Museum. In 1822 Harlan was elected to the American Philosophical Society. He published the first attempt to classify American mammals, *Fauna Americana, being a description of the mammiferous animals inhabiting North America*, and a short book, *American Herpetology*. He also translated a French treatise on embalming and appended his recommendations for cleaning flesh from skeletons. Tadpoles with their delicate "suction mouths" produced beautiful natural skeletons. Or put a skeleton in a box near an ant's nest and "these industrious operatives rapidly remove the flesh from the bones." Careless errors and a whiff of plagiarism kept the books from ever living up to Harlan's hopes, and an "unfortunate infirmity of temper" left him few friends to protect his reputation.[16]

Morton worked with Harlan in the late 1820s, helping with his dissections, illustrating his articles on whales and frogs, and drawing "the anatomical parts" of a hermaphrodite orangutan from Borneo that Harlan dissected in Philadelphia. The two collaborated well, but Harlan's truculence alienated others, and his nasty temper seems to have dimmed his original promise. He lacked Morton's gift for making and keeping friends, a skill crucial to establishing and maintaining the intellectual networks of early nineteenth-century naturalists. Contemporaries suspected that he cut corners, claiming credit he did not deserve for discoveries he had not made. In one case, he took advantage of his position in Philadelphia to rush a description of a new species of rice rat into print under his name, even though that rat's discovery and first description belonged by rights to South Carolina naturalist John Bachman. Such "duplicity," as one historian labeled it, threatened the simple trust behind a working network of naturalists and collectors. Morton benefited from Harlan's fall, and by the late 1820s collectors around the country began sending specimens to the Academy of Natural Sciences and to the attention of the more reliable Morton.[17]

But it was while the two were working together that Harlan established one of the first cabinets of anatomical specimens in the United States. Harlan thought an anatomical cabinet would boost the confidence of American naturalists, and he began to collect specimens from around the world, including more than two hundred human skulls. He lent his skeletons of a man and a horse to the Academy of Natural Sciences, but when he sailed for France in 1839, he packed up most of his specimens, storing them in a loft above a white lead factory. On a Sunday night in June, a careless watchman dropped coals from a kiln, and the building caught fire and burned to the ground. Newspapers reported that "hardly a vestige" of Harlan's collection was "found worthy of preservation." The building was insured; Harlan's collection, for which a Boston scientific society once offered $5,000, was not. When news of the fire reached him, Harlan turned around and came back to the States. He found nothing to keep him in Philadelphia and moved to New Orleans, where he died in 1843.[18]

At some earlier point, Harlan had left several skulls with Morton, although the two had parted ways by the time Morton began to collect skulls himself. Specimens Morton credited to Harlan included four Celtic skulls from the Catacombs of Paris and the skull from the battlefield at Waterloo (a favorite spot for bone hunters), the skull of a Burmese soldier, and the tattooed head of a New Zealander. Harlan's influence may have worked its way into Morton's writing, Harlan's *Fauna Americana* providing a model for the title of Morton's *Crania Americana*.[19]

Harlan also demonstrated for Morton how to make a career out of nature study in these years before the United States had a professional scientific establishment. Both men thought natural history could serve the ambitions of their young nation, providing an intellectual complement for the country's economic and political development. "What may not be expected in a country like our own?" Harlan asked. "Where the monstrous forms of superstition and authority, which tend to make ignorance *perpetual*, by setting bounds to the progress of the mind in its inquiry after physical truths, no longer bar the avenues of science; and where the liberal hand of nature has spread around us in rich profusion, the objects of our research." The country needed comparative anatomists, Harlan wrote, and "national cabinets" of specimens, although without state sponsorship, their establishment could be a challenge.[20]

Like most of their contemporaries in the community of American naturalists, Harlan and Morton began their scientific careers by studying medicine. Dr. Parrish and the University of Pennsylvania schooled them both. Both found good collaborators in Philadelphia at the Academy of Natural Sciences and the American Philosophical Society, but Harlan's eventual isolation from the gentlemanly cohort taught Morton the importance of cultivating networks of like-minded men. Friends, Morton learned, helped launch and sustain ideas, ties of friendship linking men into an emerging network of scientists. And Morton cultivated "a most winning gentleness of manner, which drew one to him as with the cords of brotherly affection," a memoirist wrote.[21]

Morton earned a degree from the University of Pennsylvania in 1820 and, thanks to a generous uncle, headed off to Europe. Like many ambitious medical men of his generation, he worried that an American degree earned little respect among European naturalists. Morton's wealthy Irish uncle offered to pay his bills at the University of Edinburgh, still the school of choice for many young men looking for careers in medicine. The university in Edinburgh rose to prominence in the second half of the eighteenth century during the heyday of the Scottish enlightenment. By the time Morton arrived in Edinburgh in the mid-1820s, the city and its university had both "sunk to rise no more," according to a critical observer. Many ambitious medical students headed to Paris or to London. Enlightenment confidence in empirical reason, however, lived on in the university curriculum and shaped Morton's approach to the world and his work.[22]

Many found Edinburgh dreary. From the outskirts, it looked picturesque, a city of compact neighborhoods. That first promise faded on closer experience of the city's dank streets. Stench from squalid sewers made walking unpleasant. City houses were packed so tightly together that "inhabitants may almost shake hands from the windows of the opposite houses," wrote a visitor, who complained that tight construction blocked sunlight. There were no parks and little greenery, since the city's shivering residents had long since burned convenient trees and bushes to fight the chill of long damp winters. Morton did not leave a record of his first impression of the city, but a young Philadelphian accustomed to America's woods and his city's scrubbed streets could find Edinburgh bleak. His fellow foreign students grumbled about the wretched inns where they boarded. Edinburgh newspapers sometimes denounced the city's students as drunken brawlers. (Morton was a studious and sober young man, and nothing in his history puts him among the rowdy students.)[23]

Despite its declining reputation, the university still drew a talented group. When Morton arrived, there were a thousand or so young men in Edinburgh studying medicine. Most were from Scotland, Ireland, and England, but the university attracted a handful of students from Europe and a few planters' sons from the West Indies. A small number also came from the United States, though Americans were scarce enough to prompt one to comment: "It is so pleasant to meet an American that I must record it." Most American students had some connection that brought them to Edinburgh—Quaker ties or old relatives. Pennsylvania's Quakers sent students from the state's medical schools, but most of Morton's American classmates came from South Carolina, Virginia, and Maryland, states where residents still felt ties to the home institutions of their Scottish ancestors.[24]

Since the university did not recognize degrees from American medical schools, students like Morton began medical studies over again with basic lectures on "Materia Medica," a course in pharmacology. Most medical students took courses on medical practice, chemistry, medical theory, and surgery. Morton also studied Latin, Italian, and French (complaining later that he had never mastered German) and attended Robert Jameson's course of lectures on geology, as he gathered skills that would suit him for his life as a gentleman, doctor, and naturalist. He skipped midwifery, which was not required for graduation. Morton went to the younger Alexander Munro's anatomy lectures, but supplemented Munro's course with anatomy lectures in Paris in the fall and winter of 1821, seasons he remembered as "the happiest of the three-and-twenty that had passed over his head."[25] Back in Scotland, he followed his classmates to off-campus dissection sessions led by the city's freelance anatomists. Dr. John Barclay, the most popular of them, sparked Morton's interest in comparative anatomy. Snuff-dipping Barclay had a reputation as a flamboyant lecturer. Students crammed his rooms, sometimes forcing him to repeat a demonstration two or three times. If a cadaver got too ripe, Barclay dipped more snuff and kept on working. Students left his demonstrations convinced of the value of dissection. Barclay also used a collection of prepared skulls to teach them to see at a glance the cranial differences among various mammals. This lesson Morton learned well.[26]

The university's zoological collections reinforced ideas from the lecture rooms. Morton had free run of the collection of specimens assembled by Professor of Natural History Robert Jameson, who profited from the university's place in Britain's imperial bureaucracy. Edinburgh alumni filled the bureaucracy's medical ranks. In fact, in the early nineteenth century, nearly a third of the doctors working for the East India Company or serving in Britain's military had studied in Edinburgh. They responded to requests for natural specimens and sent contributions from imperial outposts. Morton would have seen the skeleton of an Indian elephant, stuffed birds from the South Pacific, cases of rocks, minerals, insects, and fish, and the germ of a collection of human and animal crania that numbered about eight hundred by mid-century.[27]

Morton continued his lessons outside the classroom and spent the summer of 1822 hiking in the Alps. Glaciers on Mt. Blanc gave him a chance to combine geology with poetry, to taste a romantic's sense of the sublime. The next year he finished a thesis on pain, "De Corporis Dolore," spent a month in Ireland duck hunting and salmon fishing with his wealthy relatives, and then sailed for the States. In the fall of 1824, Morton returned to Philadelphia.[28]

Those European years shaped Morton's intellectual life. It's easy to see Edinburgh's influence in his habits of empirical observation and in his taste for comparative anatomy. As important, he learned the value of anatomical collections and the efficiency of developing collections by corresponding with bureaucrats, diplomats, soldiers, and doctors posted on imperial frontiers. But his passage through Edinburgh was too brief to take full advantage of all the university had to offer. Edinburgh held an important place on the world's intellectual map, but a quick glimpse of some of the characters who walked its streets in the 1820s suggests a complex, contested, and changing world of science.

Had Morton dawdled for a year, he could have crossed paths with young Charles Darwin, who had come to the university with his brother to study medicine. Darwin was ten years Morton's junior. He did not take to his Edinburgh medical education. He found the lectures dull and surgical demonstrations churned his stomach, biographers write. But like Morton, he enjoyed the city's cosmopolitan feel and intellectual life. Family ties opened doors to the city's scientific establishment. Darwin conversed with radicals and upstart phrenologists, listened to various preachers, collected sea creatures with his brother, learned to stuff birds, and read a lot. Morton and Darwin followed very different intellectual routes out of the city: the American carving a career that built on old arguments that species did not change, Darwin slowly gathering evidence that species were not nature's immutable constructions. (The contrast between the two men is striking. Morton supported his argument about a static world of unchanging races with his fixed and motionless skulls. It was a world as still as death, literally. The world Darwin described in *On the Origin of Species* [1859] teems with life in all its beauty, mystery, and variety, as natural selection proceeds through a constant struggle for survival. Darwin's world is never still.)[29]

American ornithologist John James Audubon, dressed in a backwoodsman's buckskin, arrived in Edinburgh in the mid-1820s too. He needed subscribers for his *Birds of America*. Edinburgh treated him well. "How proud I feel that in Edinburgh, the seat of learning, science, and solidity of judgment, I am liked and am received so kindly," he wrote. In the next decades, Audubon played a small role in Morton's skull work, checking British crania collections for him and sending him five skulls he found drying on the San Jacinto battlefield in Texas. (Here is one description: "Mexican soldier, with a cicatrised sabre wound of the os frontis. Mixed Indian and Spaniard? ætat. 30. Slain at San Jacinto, Texas. I.C. 81. J.J. Audubon. Esq.")[30]

One of Morton's most important allies, the phrenologist George Combe, also got his start in Edinburgh. All through the 1820s, Combe and colleagues in the Edinburgh Phrenological Society campaigned for their theories about human brains, lecturing, writing, and entertaining celebrity sitters, like Audubon. Audubon remembered Combe's reading. The ornithologist learned that he was generous, quick tempered, forgiving, and appreciated talent, that he had the skull of a "strong and constant lover, an affectionate father." According to Combe, if Audubon had not been a gifted painter, he would have made a good general. That idea pleased Audubon.[31]

By mid-century some critics dismissed phrenologists as scheming humbugs who pushed silly theories about character traits onto a gullible public, but in the early years of the nineteenth century, phrenology helped promote the important materialist premise that the brain was the organ of the mind. It also offered a beginning for understanding cranial location and helped establish human skulls as a favored site for investigation. During these same years, naturalists in Europe and the United States began to work on the idea that differences among the races of men could be set down as differences in the shapes and sizes of skulls. Although Combe dissented from the harsh racial implications of craniology's arguments that the differences in the shape and size of skulls expressed nature's racial differences, in the late 1830s he helped Morton finish his *Crania Americana* and tried to help sell the book in the United States and in England.

Combe will come back into this history, but a brief introduction helps fill out Morton's intellectual experience in Scotland. Combe, born in 1788, was one of twelve children of a prosperous Edinburgh brewer. His father hoped he would become a lawyer, but as a young man Combe abandoned the practice of law and took up phrenology. At the turn of the nineteenth century, phrenology swept through European intellectual circles, promising to unlock the secrets of the human brain, the physical organ that housed the mind. Breaking with orthodox metaphysics, phrenologists argued that ideas originated in the brain and not in an immaterial soul. This insight pushed some practitioners to search for mental qualities and character traits in different areas of the brain and finally to produce those skull maps that are phrenology's lasting legacy—the plaster head in the fortune-teller's window. Well through the middle years of the nineteenth century, American practitioners traveled the country, illustrating lectures with criminals skulls picked up at the gallows and casts of famous men's heads. A good phrenologist could find the sins of criminals and the accomplishments of great men written in the contours of their skulls. A reformer like Combe believed that phrenology could help teach ordinary individuals to correct their faults and cultivate their gifts.

Combe embraced the new science after he witnessed a brain dissection. Phrenology appeared to suit his ambitions to be a famous writer and help his fellow men. Affable Combe became one of phrenology's great popularizers; his *Constitution of Man*, a book with advice for readers on how to achieve true happiness, became one of the nineteenth century's best sellers. Phrenology's map of the brain would guide reformers who hoped to foster human improvement. Combe considered himself a serious naturalist, an intellectual, a reformer, a philanthropist, a man altogether different from carnival charlatans sometimes associated with phrenology.

Experience in Edinburgh gave Combe a different view and prepared him to accept phrenology's intellectual premises and social promises. Combe hoped that anatomists and moral philosophers would find common ground in phrenology and work together to investigate the physiological origins of ideas. The simple materialist notion seemed scandalous in the early nineteenth century, but scandal gave phrenology social punch in Britain's deeply orthodox intellectual community. "The phrenologists of Edinburgh must have been in the very fervor of their first love during Morton's residence there, and they included in their number some men of eminent ability and eloquence," one of Morton's eulogists wrote.[32] "Collections of prepared crania, of casts and masks became common; but they were brought together in the hope of illustrating character, not race, and were prized according to [a] fanciful hypothesis [that] could make their protuberances correspond with the distribution of intellectual faculties in a most crude and barren psychology." Historians have argued that phrenology appealed to working-class Britons, who had grown suspicious of elite learning and tired of closed university doors. They filled seats of phrenological lectures and snapped up the cheap books and pamphlets that offered intellectual expertise without university study. Some (like their American cousins bent on commercial careers) embraced what appeared to be quick ways to read the character of strangers.[33]

Morton took his Edinburgh lessons (on collecting, empiricism, anatomy, and phrenology) home to a country deeply divided along racial lines, and those racial divisions shaped his intellectual world.

When he returned to Philadelphia, Morton began to practice medicine, although he competed for patients with healers of many stamps. In the mid-1820s, Philadelphia had a population of about 100,000 (almost as large as Baltimore and Boston combined, although smaller than New York), but the city had more than its fair share of medical practitioners, including "cuppers and bleeders," leechers, midwives, steam doctors, and herb doctors. Morton's credentials from the University of Pennsylvania and the University of Edinburgh earned him a place in the city's corps of "physicians," the elite among the healers. He opened an office in his home on Mulberry Street (as that stretch of Arch Street was then known); a good address, steps from Independence Hall, the Friends'

Meeting House, and the American Philosophical Society, but a neighborhood well supplied with educated "physicians."

To make a living, Morton needed to find paying patients or get well-placed friends to appoint him to paying jobs caring for inmates, prisoners, and residents of the poorhouse. He struggled at first, finding the public "deaf" to the appeals of the young physician, wrote one friend. (Although another remembered, "It is scarce necessary for us to refer to the success of his professional, or merely medical career. That was always great.")[34] In the best of times, a successful medical practice supported his scientific avocations. But like many naturalist contemporaries, Morton worried about money. Income from his medical practice was not quite enough to support his family and feed his intellectual passions, particularly when his skull work took off.

Morton lectured on anatomy at the Pennsylvania Medical College and at the Almshouse Hospital, but over the years Philadelphia's Academy of Natural Sciences absorbed most of his intellectual energy. The academy had modest beginnings among a group of men "friendly to science" (but less friendly to religion), who began to meet in the winter of 1812 to discuss ideas about natural history—words that described their broad intellectual interests in the flora and fauna, past and present, of the New World and to pool resources to subscribe to European publications. Through the early 1810s, they met either at apothecary John Speakman's house or at Mercer's Cake Shop, a favorite spot for ice-cream-loving Philadelphians. Conversations among these gentlemen amateurs, in such informal settings, helped produce and organize new knowledge about the American environment.[35]

For the first years, the academy ran on a shoestring, without a building, a library, a garden, or a collection of specimens. Members offered specimens, books, and finally a building to the country's only institution dedicated to natural history, but finances remained uncertain. In 1817 William Maclure, a wealthy Scots-born merchant and amateur geologist, decided to underwrite the academy, funding its journal, purchasing mineral specimens, and donating a collection of scientific books published in Revolutionary France. Maclure's donation gave the academy the core of a library whose books offered tacit lessons on Republican science and ideas about the role that natural history played in educating virtuous citizens. In return, colleagues elected Maclure the academy's president.[36]

By the mid-1820s, the academy had collections of crystals, rocks, insects, shells, and birds, along with fossils that New Jersey farmers dug from the marl and the material Lewis and Clark had collected on their journey across the continent to the Pacific. Even if they never left Philadelphia, academy members began to imagine a building to house the specimens that friends in the field continued to send their way. Theirs would be one more institution built on the foundations of local curiosity, yet another reason to imagine Philadelphia as America's Paris. Botanist John Bartram had left the city the beginnings of a botanical garden, and Charles Willson Peale had given Philadelphia the country's best museum, with a fine collection of stuffed birds, mammals, and fish and the prized skeleton of the great mastodon.[37]

The tasks of running the academy fell to Morton in 1825 when Maclure and several colleagues left Philadelphia to join Robert Dale Owen's socialist utopia in New Harmony, Indiana. It seemed an eccentric decision even to contemporaries, who worried that Maclure had been "humbugged" by Owen's enthusiasm, but Maclure convinced zoologist Thomas Say, French artist-naturalist Charles-Alexandre Lesueur, and some of the more radical members of the academy that America's future lay in small settlements like Owen's and not in its growing cities like Philadelphia.[38]

Quiet and steady Morton thrived in the vacuum they left and threw himself into the academy's work, managing the correspondence that assured a steady flow of new specimens from amateur naturalists around the country. He directed Academy of Natural Sciences publications, too, and cultivated an analytic expertise on subjects ranging from geology and fossil vertebrates, to "Parasitic Worms," albino raccoons, and a "supposed new species of Hippopotamus." When Morton decided to collect crania—skulls of animals and humans—he knew where to turn for specimens and how to present his ideas to an international community of naturalists who had come to value careful analysis in the study over fleeting observations in the field. Morton's intellectual and institutional work might have come down to us as a legacy of a hardworking Philadelphia naturalist had he continued to work on worms or raccoons or subscribed to the radical egalitarianism of the academy colleagues who had followed Maclure to Indiana. But his work on differences among men changed that. Morton's training as a naturalist gave him the discipline, language, and skills to analyze variation in the shape and size of human crania. Morton contended that his work involved "discovery" of the facts of nature's racial differences, but it is more useful to explore his role—not in the discovery of those differences, but in their invention. In the 1850s, Morton's work found a receptive audience among slaveholders and their supporters ready to accept a naturalist's account of racial differences.[39]

With his medical practice, his family, and his intellectual obligations, Morton rarely ventured far from Philadelphia. Since specimens, things to study, came to him, he did not have to leave the city to sharpen his naturalist skills. In fact, his city perch gave him an intellectual advantage, as institutions like his Academy of Natural Sciences weighed in on the significance of specimens and observations sent from correspondents scattered around the country and around the world.[40] Morton's sedentary habits make the one account he left of a rare trip outside the city all the more striking. In the spring of 1834, he accompanied an elderly patient to the West Indies. His descriptions suggest some of the ways a naturalist's observations shaded into his social opinions. Morton and his patient traveled through Barbados, St. Lucia, Martinique, and Dominica, adding quick stops in Guadeloupe, Antigua, Montserrat, Nevis, St. Kitts, St. Eustatius, Saba, Tortola, and St. Thomas. The record he left of that trip betrays the mean streak in the racism that came easily to men like Morton, warping his naturalist's observations into social commentary. "In my eyes," he wrote, the Africans of Barbados have a "very repulsive appearance." In his eyes, the women were "thin and squalid"; and to his ears, their voices sounded shrill.

Such observations gave an empirical foundation for what came next. These people, those enslaved and those recently emancipated, were "sulky," working only under an overseer's whip. "The philanthropist" argued that slavery explained the terrible conditions, Morton acknowledged. But for naturalist Morton, it worked the other way: "When freedom gives them the choice between industry & idleness, a vast number of them will prefer the latter condition, even at the expense of penury, hunger, and disease," he wrote. Natural indolence brought the lash down on African backs. Slavery's defenders wove observations like these into justifications for their peculiar institution. Morton did not go quite that far, observing that "gradual emancipation, like that adopted in some of the United States, would have been conducive equally to the happiness of the Negro, and much more to the security of the master and the prosperity of the colonies." Yet his tone makes it easy to see why some slaveholders found him a good ally.[41]

". . . and Ethnologist"

Morton modeled his skull collection on the anatomical cabinets he had seen in Edinburgh and Philadelphia (and on collections he had read about), but concentration on human skulls, particularly the skulls of Native Americans, distinguished his collecting. Morton took up questions that comparative anatomists had asked about the shape and size of skulls of different animals, but instead of looking at various animals, he compared human races. His way of thinking highlighted small differences in the size and shape of skulls and differences in the average internal capacities of different groups. Morton acknowledged that if he pursued his investigation of difference to its logical end, he would have to argue that humans of different races represented different species of humanity. Morton subscribed to a common view that "species" expressed the original differences of divine creation and did not change.

He hinted that human races might represent different species, even as he advanced his arguments over shaky ground. His phrasing was tentative as he approached controversial questions about descent and tried to square his belief that differences among species preserved the original differences with the biblical story of Adam and Eve. "Without attempting to pursue this intricate question in detail," he wrote at the start of *Crania Americana*, "we may inquire, whether it is not more consistent with the known government of the universe to suppose, that the same Omnipotence that created man, would adapt him at once to the physical as well as to the moral circumstances in which he was to dwell upon the earth? . . . [W]e are left to the reasonable conclusion, that each Race was adapted from the beginning to its peculiar local destination. In other words, it is assumed, that the physical characteristics which distinguish the different Races, are independent of external causes." The conclusion that looked reasonable to Morton left others puzzled. Questions about the accuracy of the biblical account of creation had plagued thinking people since the fifteenth century as accounts proliferated of humanity's astounding diversity.[42]

But Morton had other troubles. In his intellectual world, male and female of different species could breed, but could not produce fertile offspring. If human races were indeed different species, how was one to account for the obviously fertile offspring produced by sex between humans of different races? Morton speculated that offspring were less fertile, but he also considered abandoning interfertility as the criteria to distinguish a species. As he wrote in an 1847 essay, "Since various different species are capable of producing together prolific hybrid offspring, hybridity ceases to be a test of specific affiliation."[43]

He constructed his arguments about biological difference out of cultural materials, which operated by their own particular logics. Morton capitalized on decades of enthusiasm for taxonomy—for conversations about the order of nature that extended from learned men to ordinary folk, from naturalists to farmers and fishermen. But he turned those conversations to racial difference. And work with skulls gave his ideas an extra dose of authority. Anatomists had long compared animal skulls, but until Morton came along, no naturalist or anatomist had human skulls in anywhere near a number that approached his. Although war and removal facilitated collecting Native American skulls, Morton and his collectors relished the challenges in their work. "The Indians have an extraordinary veneration for their dead," Morton wrote, "which sometimes induces them, on removing from one section of the country to another, to disinter the remains of their deceased relatives, and bear them to the new home of the tribe."[44]

Morton had forebears in the skull business: close to home, his grumpy Philadelphia colleague Richard Harlan; further afield, the Dutch artist/anatomist Peter Camper (1722–1789) and the German naturalist and comparative anatomist Johann Friedrich Blumenbach (1752–1840). Each worked with skulls, but none had Morton's success in filling a cabinet with crania. Camper, who sketched the well-known scheme that sets a sculpted bust of the Greek god Apollo as the ideal human profile, got Morton thinking about the shape of skulls. Blumenbach, professor of anatomy at the University of Göttingen, convinced Morton that skulls preserved distinctive characteristics of humanity's five races—the Caucasian, the Ethiopian, the Mongolian, the American, and the Malay. But Morton thought of comparing the internal capacity of the skulls of various races.[45]

Life on Amsterdam's busy streets inspired Camper's sketch of the perfect profile. By the middle years of the eighteenth century, people from around the world passed through Amsterdam, and to Camper the city seemed a "concourse of people assembled together . . . from the different quarters of the globe." Some Amsterdam residents had honed visual skills to "distinguish, by a single glance of the eye, not merely negroes from white men,—but among the latter we discriminate Jews

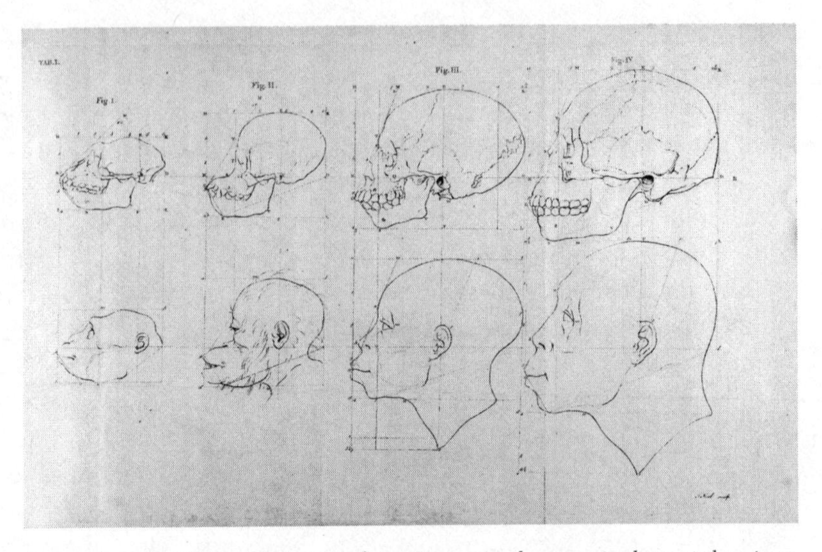

Fig. 3. From *The Works of the Late Professor Camper, on the connexion between the science of anatomy and the arts of drawing, painting, statuary, &c., &c.* (London, 1821). (Courtesy of the New York Academy of Medicine.)

from Christians, Spaniards from Frenchmen or Germans, and these from the English." Camper wondered how faces carried these national identities. Training as an anatomist led him to look for explanations beneath the surface. He hoped to help his artist friends, who, though able to distinguish faces in crowds, had no clear ideas about the source or significance of the differences they saw. Their ignorance harmed their portraits. Most painted Africans as tinted white men. ("Un blanc barbouillé de suie," agreed French anatomist Georges Cuvier.) Artists saw skin color but ignored the facial angles written in the bones of human skulls. Camper thought these angles distinguished human races. The discovery "formed the basis of my edifice," he wrote.[46]

Camper's discovery has come down to us as one of the principles of racial ordering meant to express Caucasian superiority. He based his insight on a small collection of human crania, gathered from eight "nations." He had a few Dutch skulls, "two of English negroes (the one a young person, the other advanced in years)," a female Hottentot, a youth from Madagascar, a man from the Celebes, and the cranium of a Calmuck from Mongolia. He lined them up beside the skull of an orang-utan, and in a sequence he found "amusing to contemplate" discovered that his "divers European" skulls had profiles closest to the god Apollo, the negroes closest to the ape. Camper's amusing scheme became one of the iconic images of racial science.[47]

And the image derived its power in part from its simple depiction of skulls. Skulls, the "receptacle of the brain, of the organs of the senses and the masticatory apparatus," appealed to men ready to celebrate the power of reason and observation. Skulls expressed humanity, generally, and "race-characters more striking and distinguishing than those presented by any other part of the bony system," Morton's collaborator James Aitken Meigs wrote in the 1850s. But work with skulls also reinforced a social position for naturalists, who had begun to distinguish their work as disciplined or objective inquiry.[48]

The difficulty of getting skulls made them all the better to work with. Camper wondered how to gather a representative sample of skulls from the world's diverse populations when sailors refused to bring human heads back to Europe. Amsterdam's concourse of people gave him a glimpse of the enormous range of human variety. On occasion, one of these wanderers would die in the Dutch city. Camper said he had once seen a Chinese man on an Amsterdam street, and he acquired the skull of another who died in Europe. But he had never been able "to obtain possession of the cranium of a Native American, nor even an Anglo-American." (A meeting in London with the Pennsylvania-born artist Benjamin West sparked Camper's curiosity about Anglo-Americans. West suspected that he and his fellow colonists had begun to develop distinctive traits—skull shapes and facial features that a perceptive artist would discern.)[49]

One skull sits at the heart of Camper's visual scheme. Like many collected skulls, this one records the trace of a story of someone who died far from home, far from the relatives who would have disposed of a dead body. Camper traced the African profile in his scheme from "the cranium of the young negro," an eleven-year-old from "Angola," just "changing his teeth," a touching note that keeps that boy forever at his first molars. Camper dissected the boy's body "publicly, at Amsterdam, in the year 1758." It's a good guess that the boy had come to Amsterdam as the servant or slave of a European trader and that the pair mixed into the city's cosmopolitan crowd. But when Camper sketched this boy with his new molars, he gave him a permanent role in the intellectual history of European racism. Camper reported that he found no difference between this boy's body and brain and the body and brain of a European boy of the same age. But the angle of the African boy's profile did differ, slanting for Camper halfway between the skulls of apes and the skulls of men.

Blumenbach also made aesthetic claims about the skulls he collected. The German began his inquiry into the natural history of man by working on taxonomist Carolus Linnaeus's description of humans suited to the continents of America, Europe, Asia, and Africa. He asked the students he had trained at Göttingen to please send him skulls. These educated men proved a more reliable source than superstitious sailors, and with their help Blumenbach established Europe's first significant scientific collection of human crania, with some 250 specimens from around the world. He concluded from his work with skulls, "*That no doubt can any longer remain but that we are, with great probability, right in referring, all and singular as many varieties of man as are at present known, to one and the same species.*" (Morton, who studied Blumenbach's published *Decas collectionis suae craniorium*, came to a different conclusion about the human species.) Blumenbach, nevertheless, confessed to preferences among his skulls. He said that every visitor to his collection found the skull of a young Georgian woman his most beautiful specimen. The skull's symmetry pushed him to an odd imaginative leap, and he named her the representative of humanity's original form, an ancestor of the group he named with a new word, the "Caucasian" race. Europe's contemporary population had descended from original Caucasian stock.[50]

Camper's scant experience with skulls hardly supported his generalizations. And even Blumenbach's two hundred skulls were too few to make valid comparisons among human groups. Both acknowledged the challenge of getting skulls from faraway places. Superstition about dead bodies explained the problem, but so did an intellectual reticence to push too far into the natural history of humanity. Comparative anatomist Georges Cuvier (1769–1832) addressed the skull-collecting problem directly, writing out instructions for a one-eyed medical student bound for the Pacific. The young naturalist studied Cuvier's instructions daily, he said. Cuvier told him to look for skulls where people buried the dead or where Europeans had fought native peoples and scattered them with musket fire before they could carry off their dead. He recommended soda or caustic potash to clean the bones or, if possible, a solution of a corrosive sublimate to preserve the flesh. Preserved heads, specimens with facial features in tact, would be real treasures. But transporting human remains could be difficult, since many sailors believed

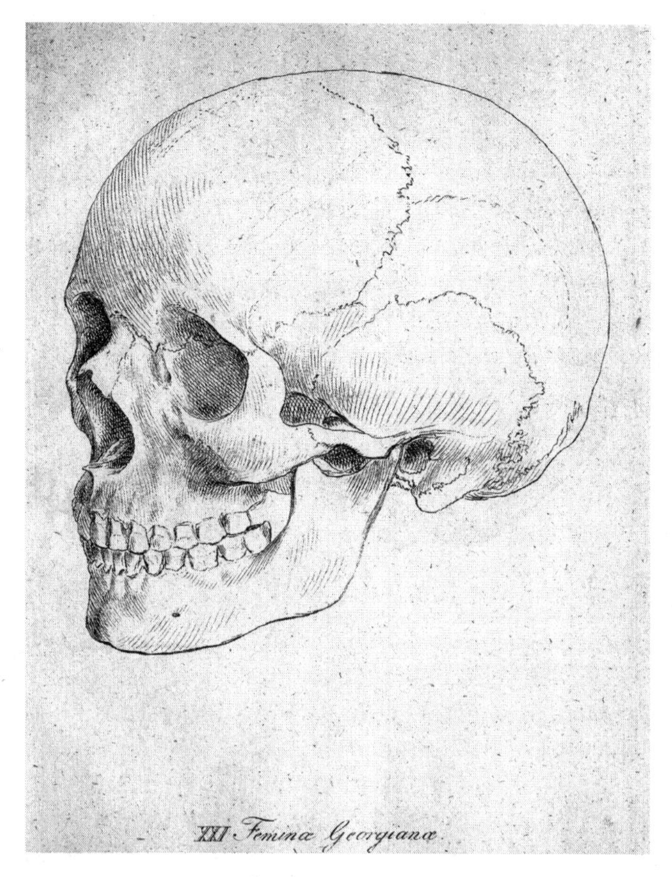

Fig. 4. "Feminæ Georgianæ," from Johann Friedrich Blumenbach, *Nova pentas collectionis suae craniorum diversarum gentium* (1828), plate XXI. (General Research Division, The New York Public Library, Astor, Lenox and Tilden Foundations.)

that a cargo of corpses doomed a ship in a storm. Cuvier encouraged his protégé to face down less-educated critics and to remember that reason, not superstition, governed the work of naturalists.[51]

Though Blumenbach's work influenced Morton, conclusions about racial difference played differently in the social and racial circumstances of the mid-nineteenth-century United States. Morton acknowledged that even if his country's literary output lagged, "opportunities for anthropological research" abounded in the United States. And that research gave American naturalists the opportunity to bring distinctive information to conversations among naturalists in Europe. One of Morton's eulogists also thought his work had practical value for the "management" of men. As he wrote, in the United States "three of the five races, into which Blumenbach divided mankind, are brought together to determine the problem of their destiny as they best may, while Chinese immigration to California and the proposed importation of Coolie laborers threaten to bring us into equally intimate contact with the fourth." "It is manifest that our relation to and management of these people must depend, in great measure, upon their race-character."[52] A messy mix of the practical and abstract, of the low and lofty, of the aesthetic appeal of beautiful crania and history's sad cases of unburied dead, begins to explain the appeal of Morton's project. He had intellectual aspirations, multiple sources for human skulls, and audiences interested in his work.

America's Golgotha

When he decided to collect skulls, Morton reached into the network of far-flung naturalists he had established during his years at the Academy of Natural Sciences. Amateur naturalists around the country sent the Philadelphia institution specimens—rocks, plants, fossils, birds, insects—and observations. And Morton and his city-dwelling friends rewarded their countrymen by commenting on their donations and ideas. "In looking over his correspondence now," a friend wrote shortly after Morton's death, "it is surprising to see the number of men, so different one from another in every respect, who in all quarters of the globe were laboring without expectation of reward to secure a cranium for Morton."

His 138 donors included missionaries in Africa, doctors in Florida and Cuba, diplomats in Mexico and Cairo, white settlers sulking through hot summers in Indiana, soldiers in Georgia, explorers in the Arctic, scientists in Oregon, and a president of Venezuela. They remembered Morton on their summer tramps and expeditions and whenever they chanced on dead bodies. On his way to Oregon in 1835, ornithologist John Kirk Townsend "often thought of my friend Dr. M. and his *golgotha*," as he kicked buffalo skulls bleaching on the Kansas prairie. As they collected skulls, Morton's friends teased him about his Golgotha, his study that had become a "place of skulls." But those friends also helped him build his Golgotha. Morton cataloged "22 varieties of the

Caucasian race; two varieties of the Mongolian race; 13 of the Malay race; 69 of the aboriginal American race; 21 of the negro race; 8 of the mixed races; besides skulls of lunatics and idiots of several races."[53]

Morton's requests sent amateurs looking for skulls—animal as well as human. Here is a typical request from Morton: "If you can obtain skulls of any animals from *Homo Sapiens* to the lowest link, I wish you to take care of them; by making inquiry you may possibly meet with some Indian Crania. I should also be much gratified to possess an entire skull of the Florida *Manatee*," he wrote an acquaintance in 1832.[54]

Or as Morton wrote to a missionary in Liberia: "I am endeavoring to make an extensive series of this kind, which already embraces about 500 specimens, but as yet there are no African subjects among them. The heads of your quadrupeds, Birds, and reptiles, and even fishes will be very important to me."[55] Five years later, Morton's friend S. M. E. Goheen responded. He sent him the skull of a hippopotamus, the skulls of an Eboe man and woman hanged for murder, and the heads of five others killed in 1840 in an attack on American settlements around the Methodist mission at Heddington. The donation included the skull of man named Gotorah. American settlers had killed him and stuck his head on a post outside the mission to scare back the Africans who were trying to keep the newcomers off their lands. One settler "reserved it for the Governor, with Greegrees, a great quantity, which I delivered to the Governor." Morton cataloged the skull as specimen "**1093**, Golah Negro, warrior, ætat. 70. F.A. 77° I.C. 85."[56]

Letters also capture the pleasure that collecting specimens and speculating on nature could give to daily lives. Here, for example, is an Indiana man worrying through a steamy afternoon about an "Epidemic Fever" that seemed to be killing "our common house fly. I say fever for they collect where they can procure cold water and drink with avidity until the Abdominal Parieties give way and they cohere to the spot where they have been drinking and continue so until death." Could his dying flies be important? A sign of a coming plague or a plague among insects?[57]

Family obligations and poor health—a bout with pneumonia that left him with only one working lung—kept Morton in Philadelphia, but he played confinement to his intellectual advantage, honing the analytic abilities that let him cast himself as superior to the field naturalists who could collect specimens but could not approach Morton's capacity for contemplation. Work comparing skulls led Morton to two conclusions, although they were not universally shared. First, that polygenism—the idea that God must have created more than one species of man—offered the best explanation for the differences among people. And second, that of the five races or species of men, Caucasians had the largest skulls and, therefore, the biggest brains.[58]

By the end of his life, Morton had skulls from every continent, although his first specimens came from local institutions for the poor: "two Negroes, born in the United States," a "Negro idiot," and a boy who suffered from hydrocephaly. Next came the skull of an Anglo-American woman who had died near her hundredth birthday. Morton set out to describe physiological differences in these skulls, but he could not have missed the simple social fact that united them. No family members had claimed these corpses, and like many who died in similar situations, their bodies went off to medical schools, labs, and cabinets, like Morton's. He learned from his earliest specimens that idiocy and poverty produced skulls.

Acquaintances, like diplomat and showman Marmaduke Burrough, helped expand the collection's range. Burrough donated five skulls from India—skulls of two "Bengalee children," two "Bengalee men," and a "Young Hindu Woman, *burned with her husband near Calcutta*."[59] However interesting these specimens, Burrough helped give the collection its real direction with the donation of the skull of an ancient "Chimuyan" from Peru, Morton's first Native American specimen. Picking up the American theme, Morton bought the skull of a Huron chief from the "estate of a German traveler" and noted the following: the skull had belonged to a man "about 60," "killed near Detroit in a *rencontre* with another Indian," or as he phrased it later, "slain in a broil with his son-in-law."[60]

Morton's collectors haunted executioners' grounds as well, but he milked this old source of bodies for science for specimens of races rare in Philadelphia. He believed he had skulls of Chinese pirates hanged at Singapore, an Englishman executed for piracy and murder in Philadelphia, and an "Anglo Saxon" hanged for murder and cannibalism in Australia. This last would be big-headed Alexander Pierce, a nasty man but with an impressive cranial capacity of 99 cubic centimeters (well above the English mean of 96) that made him a very valuable recruit for the Anglo-Saxon column on comparative skull capacity.

But any self-respecting curator would question the shaky provenance of the cannibal skull. Morton's friend Marmaduke Burrough told him that he received the skull from a "W. Harris of Calcutta," although he was not sure how the skull found its way to India from the prison colony in New South Wales. (Harris admitted to having read an account of the criminal in a Philadelphia paper, "which found its way to Calcutta." Burrough might have asked Harris when that story was attached to this particular skull.) Morton was happy to add the skull and its story to his collection. Pierce was a sailor, born in Scotland or Ireland. It is not clear what crime landed him in Britain's penal colony in New South Wales, but he earned his lasting notoriety when he organized a prison break with a band of comrades. Supplies ran short, and hungry Pierce killed his friends "and after ate their flesh." Pierce was found, tried, and hanged for murder. A doctor at Hobart dissected his body. And then his skull began its last long trip, through many hands, and around the world. Or so Morton wanted to believe. Descendants and tribal members have reclaimed many of Morton's specimens, but Pierce's large skull still sits in a drawer in the basement of the Museum of Archaeology and Anthropology at the University of Pennsylvania. No relative has come for man-eating Pierce.[61]

Skull collecting became more common in Morton's generation, but it did not become commonplace. The challenge was to give shape to a collection that came together by chance and accident. Could travelers help? The American explorer and diplomat John Lloyd Stephens promised Morton he would look for Mayan skulls on his travels through Central America. But bone collecting disturbed him. "In fact, alone in the stillness and silence of the place, something of a superstitious feeling came over me about disturbing the bones of the dead and robbing a graveyard. I should nevertheless, perhaps, have taken up two skulls at random, but, to increase my wavering feeling, I saw two Indian women peeping at me through the trees, and, not wishing to run the risk of creating a disturbance on the hacienda, I left the graveyard with empty hands," Stephens wrote.

Still, he did not forget his promise to Morton. When Stephens's cook died, he imagined her among the remains displayed in Morton's cabinet. "Alas! Poor Chaipa Chi, the white man's friend! never again will she make tortillas for the Ingleses in Uxmal! A month afterward she was borne to the *campo santo* of the hacienda. The sun and rain are beating upon her grave. Her bones will soon bleach on the rude charnel pile, and her skull may perhaps one day, by the hands of some unscrupulous traveler, be conveyed to Doctor S. G. Morton of Philadelphia."[62]

Morton wrote polite letters as he sought to expand his donor base and cultivate new sources for the specimens he needed. He adopted a gentle tone when he approached Dr. José María Vargas, the president of Venezuela, to congratulate him on his election to honorary membership in the Academy of Natural Sciences. Vargas had trained as an anatomist, but Morton acknowledged that affairs of state might now cut into his scientific endeavors. He explained that he had been "engaged for years past in making a collection of the Crania of all Animals from Man to Beasts, Birds, & Reptiles." Yet he still did not have enough skulls to represent "the Aborigines of the Mexican and South American provinces," in a forthcoming "Crania Americana; or a comparative view of the skulls of the aboriginal inhabitants of North and South America." "Although I cannot for a moment expect a gentleman of your pressing & responsible avocations to render me any assistance in this question, yet I trust, as before mentioned, that some one of your acquaintances may not consider it beneath his notice." Vargas's stay in the presidency was brief, and he found time to help Morton, who, in his final accounting, thanked Vargas for three skulls, including the "skull of a Charib of Venezuela, flattened by art," the skull of a man, who had died at forty. His remains had been "found in a terra cotta vessel, wherein it had probably been preserved for centuries." And the skull of "Sambo: mixed race of Venezuela Indian and Negro: man ætat. 40. I.C. 81. Ex President Vargas, of Caraccas [*sic*]."[63]

Morton had to rely on the credentials of collectors to vouch for the authenticity of the specimens they sent. But fraud stalked the skull trade, and Morton's grandiose desires made him an easy target. He received four Araucanian skulls "through the kindness of Dr. J. N. Casanova of Valparaiso, who informed me that the three heads, Nos. 654, 655, 656, were taken from chiefs killed in an encounter with the Chilian army under General Bulnes, on the river Bio-Bio, in 1835." Contemporary history lent them particular value. But on a closer look he discovered scraps of mummy wrapping, and Morton concluded that "these were not *recent* crania." "Dr. Casanova, however, could not suppose that he had been deceived by his agent, and I therefore published the circumstances as related by him, and on his authority, in my Crania Americana, page 243." Worse, it now appeared to Morton that "judging from the size and conformity of the skull No. 654 . . . that it had belonged not to a chief, as was supposed by Dr. C., but to a woman," and in the annals of collecting, that meant Morton had been tricked into buying a lesser specimen.[64]

More common complaints swirled around the price on heads. Morton made a virtue of his personal investment in his project, modeling what he imagined as the democratic science of dedicated amateurs. "Dr. Morton began alone," a colleague remembered, "and with nothing; without patronage of government, or the assistance of imperial or royal treasuries: . . . yet by means of his own pecuniary resources, never superabounding; overwhelmed with professional business; offtimes [*sic*] in miserable health and in danger of death, he had, so far back as 1840, collected and arranged a cabinet of 867 human crania, from many widely separated regions of the earth, 253 crania of mammals; of birds 267, and of reptiles and fishes 81—making 1468 specimens."[65]

Most collectors donated crania, happy to be promised recognition in Morton's publications. But that was not always the case, and one exchange gives us a glimpse of Morton the bargainer. In the mid-1830s, a Dutch doctor and naturalist named Jacob Elisha Doornik (1777–1837) offered to sell Morton a collection of Malay skulls. Doornik had collected Malay skulls when the East India Company sent him to Batavia to inspect hospitals. He picked up skulls in the hospital trash, from executioners, and even along riverbanks. As another traveler wrote, up-country violence and fishermen's accidents left the coast awash in skulls. "In going up the canal from a ship lying in the harbour, you are certain to meet large quantities of putrid animal matters floating down: what with the sight of dead Malays, in every stage of putrefaction, and torn in pieces by the alligators."[66]

Doornik saw an opportunity to turn Batavia's waste into money in America, but shaky English may have checked his plans. He landed in the States sometime in the late 1820s and began writing to Morton in French, sharing theories about organic change he thought he saw in fossils. He promised he was working hard on his English because he wanted to teach American audiences about the Malay. A crate of skulls and a head filled with stories crafted out of his experiences on the edge of the Dutch empire should have been material enough for lectures on comparative ethnology, but he could not find American audiences interested in paying to see his skulls or listen to his adventures. Doornik's confidence in American curiosity is understandable, since the ticket-buying public supported dozens of touring lecturers in the 1820s and 1830s. In the century's middle decades, successful lecturers included clergymen, college professors, phrenologists, explorers, and celebrated intellectuals, like Ralph Waldo Emerson and Oliver Wendell

Holmes. Audiences flocked to spectacular displays on chemistry and electricity. Doornik's skulls must have seemed dull fare in this company, and he had neither fame nor local connections to help him along. His Dutch-accented English may have made him hard to understand. He should have listened to Morton's naturalist friend Charles Pickering, who explained that "the American people seem to prefer a well-got-up hoax, an ingenious lie, to any truth you can tell them, however important. A lecturer in this country must be very careful not to propose to *instruct* his audience."[67]

As he struggled his way into a new intellectual life in the States, Doornik likely heard rumors that Morton was in the market for Malay and Mongolian skulls. He offered Morton his skulls at a bargain price because, he said, "I am but a naturalized citizen of the U.S. but I love this part of America as my adopted fatherland." He explained that he would use money from the skulls to buy specimens to illustrate a new series of lectures on geology, a subject sure to draw American audiences anxious to explore nature's deepest secrets about the age of the earth and the origin of life. Doornik was ready to retool, but no one would give him rocks or fossils on credit. As he told Morton, "I am a stranger in this city and can for this reason not expect any credit neither ask for it."[68] Would Morton's community of naturalists help support this patriotic stranger? Doornik thought his Malay skulls were worth at least $700 (a pricy $15,000 today), but he offered them to Morton for $250. Too much, Morton said. But Doornik packed the skulls off to Philadelphia anyway, annoying Morton, who refused to be tricked into paying Doornik's asking price. He offered him $70.

Something had gone wrong in the customary patterns of polite exchange among naturalists. Doornik checked his temper, reminding Morton that the correspondence "on the purchase of this collection of skulls has been from the beginning in a friendly way, and so I desire it shall end." He accepted the $70 because he wanted to encourage Morton. He flattered Morton, describing him as heir to the intellectual traditions Benjamin Franklin had established in Philadelphia and assuring him that his work would bring honor to his city and his nation. Morton's tone softened. "I feel much indebted for your kindness in parting with this collection, which could not have been obtained in this country thro any other sources." And then he promised that "if I live to publish my *Liber craniorium*, I shall dedicate the volume to a few friends who have chiefly aided me, & to yourself among them."

Morton did not keep the promise to dedicate his book to Doornik, although his last catalog of skulls acknowledged the Dutchman's "donation" of ten Malay skulls, the skulls of an "Oceanic Negro, from the Indian Archipelago," a "Malay Idiot of Amboynia," a Chinese man hanged for forgery, and "a convivial and dissipated Dutchman of noble family, dead at age 30." (Morton concluded that this "Dutch gentleman," whose skull had a capacity of 114 cubic inches, had had the largest brain of those whose heads he had collected.) The catalog stages a small tableau of Dutch colonial enterprise in the Pacific and records Doornik's fate—a Dutch-born doctor travels the world and dies in New Orleans in 1837.[69]

The Afterlife of Morton's Skulls

So here is Morton with his assembled skulls, representatives of the Caucasian, the Mongol, the Malay, the Ethiopian, and the American races—the remains of cannibals, criminals, chiefs, warriors, rebels, and poor children all gathered together unburied in Philadelphia. Morton did not die a rich man—his collection of crania was his principal asset. Forty-two friends raised $4,000 (a fraction of the $15,000 Morton estimated the skulls had cost him) to buy the collection from the estate and donate it to the Academy of Natural Sciences.[70] For a generation, "these vestiges of humanity" grinned at academy visitors from the shelves of eighteen windowed cabinets, Dr. Sanford Hunt wrote in a tribute to Morton. The crania drew visitors, "free of charge, on the afternoons of Tuesdays and Saturdays throughout the year." Some took a more temperate view. Morton's memoirist quoted a letter from an unnamed "eminent British ethnologist." He thanked the Philadelphia naturalists for "the privilege of even reading the catalogue of such a collection, and adding that he would visit it anywhere in Europe, although he cannot dare the ocean for it."[71]

Skull donations continued at the academy through the decade after Morton's death, bringing the count of human crania to 1,225. In 1856, for example, the academy received Eskimo skulls that Dr. Elisha Kent Kane had "taken" on his celebrated, but ill-starred, search for the arctic explorer Sir John Franklin. In 1857 navy doctor Benjamin Vreeland gave the academy another seven skulls of Greenland Eskimo and two rare Japanese crania from "Loo Choo," the open city we know as Okinawa.[72]

In 1860, as the country moved toward war over the ownership of living bodies, one of Morton's early biographers summarized the lessons Morton had extracted from the dead. "This, then, is the doctrine. Each of the pure, unmixed races has a cranial capacity and form, which is one of its most marked and permanent conditions. In a word, there is a permanent inequality in the size of the brain of different races of men, and also a variety of shape and contour of the braincase, which is almost equally marked and descriptive." Equality was a sentimental luxury, he wrote. Sober science accepted Morton's observations as "the records of that Providence which dictates the pages of human history."[73]

As we shall see, his ideas took on a life of their own, even though popular interest in his skulls waned. Curators tried to preserve the collections, but mildew ate into some skulls and others went missing, apparently borrowed but never returned by scholars and lecturers. Students curious about craniology stopped in to consult the collection. Some had ideas about human history radically different from Morton's. The Canadian archaeologist Daniel Wilson came to the academy with his theory that culture, not nature, shaped skulls. Ask any hatter, Wilson wrote, no man's head is perfectly round. "Humankind is everywhere and in all ages the same and . . . variations in culture and attainments were due to the circumstances in which people were placed and not innate racial character," Wilson wrote.[74] Others came with specific anthropological or archaeological questions. Did a skull unearthed in the ruins of Chaco Canyon match any of those in Morton's collection? Anthropologist Franz Boas borrowed skulls from the academy in the 1890s to help support arguments like Wilson's that skull shape changed as culture changed.[75]

In 1892 Morton's collection had one last moment of fame, when the academy chose forty-four of Morton's Native American skulls to be part of the United States exhibition at Spain's commemoration of the four hundredth anniversary of Columbus's voyage to the New World. The Madrid exposition had two purposes: to display what had been preserved of the Americas as first encountered by European explorers and to trace some of America's first influences on civilization in Europe. Institutions in the United States sent collections of stone axes, awls, spears, and arrows, along with belts, pipes and beads, forty-four skulls from Morton's collection, and a copy of his "great work" *Crania Americana*. By the 1890s, scholars had begun to sort humans by language, not skull shape and size (which had proved fickle markers of difference in these closely related groups), to trace ancient American affiliations, yet exhibition judges nevertheless awarded Morton's skulls a silver medal, third prize in the exhibition's system.[76]

When the skulls arrived back in Philadelphia, curators put them in storage. Years passed and the dust yellowed the varnish on Morton's skulls. At the turn of the twentieth century, Aleš Hrdlička, America's leading physical anthropologist, dismissed the collection as a worthless remnant of misguided phrenology. Still, the academy held on to the skulls. It is never easy for a museum to part with collections, and human remains can be particularly difficult. In 1938 an industrious academy employee took one of the old catalogs and tried to reassemble Morton's collection of crania. Frustration slipped onto the typed note cards, an emphatic finger on the question mark: "What has become of this utterly irreplaceable material???" What was left of the collection went to the University of Pennsylvania's Museum of Archaeology and Anthropology in the mid-1960s when the academy got out of the human remains business, as one curator put it. The skulls did not stay long at the museum. During the last two decades, descendants and activists determined to rebury the dead have taken many of Morton's old skulls, leaving only a lonely remnant of picked-over crania.[77]

574. Indian of the KALAPOOYAH tribe of Oregon: artificially compressed. Man, ætat. 50. F.A. 68°. I.C. 91.

—J. AITKEN MEIGS, *Catalogue of Human Crania, in the Collection of the Academy of Natural Sciences of Philadelphia* (1857)

CHAPTER 2

A Native among the Headhunters

Philadelphia, December 1838

Winter wind sent temperatures into the twenties, but Philadelphia's naturalists braved the cold and went out to meet William Brooks, a young Chinook man with an artificially flattened skull. Brooks had traveled east from the Oregon coast with Methodist minister Jason Lee and was working with Lee to raise money for the fledgling Methodist mission in Oregon's Willamette Valley. The trip brought them to Philadelphia late in December. They stayed with Dr. William Blanding, a mission supporter and amateur herpetologist. Blanding introduced the pair to his naturalist friends who came to see this young man's flat head. Brooks's mother, like mothers from many tribes west of the Rocky Mountains, had strapped the newborn between two boards and let the pressure mold his soft skull. Her grown son still carried marks of her handiwork. Brooks had grown up in a world where children who had been nursed by good mothers all had skulls like his. But in the eastern United States, he stood out as a curiosity among a round-headed people. Curiosity drew American crowds—scientifically minded naturalists, pious Christians, and ordinary folks anxious to see a flat-headed man from Oregon.[1]

During the winter months of 1838–39, Lee introduced Brooks to Methodist congregations around the northeastern United States. Working together, the men raised money from meetings with farmers in small towns in northern Vermont, factory workers in Lynn, Massachusetts, congressmen in Washington, D.C., and naturalists in Philadelphia. Churchgoers opened their purses when they heard Lee describe his mission work and listened to William Brooks talk about his Christian faith. Methodist records describe Brooks as a poised and gracious young man, probably about nineteen. He spoke and read English and answered questions about his soul and his body. The eastern public clamored and stared, and many who heard Brooks speak confessed curiosity about his head as well as his soul. Could a flat-headed man be a true Christian? Bostonians, anxious to know, bought tickets ("Price, 25 cents") and packed the Bromfield Street church. We sometimes imagine that Americans in those years cut clear lines between piety and entertainment, but missionary Lee and convert Brooks knew better.[2]

Philadelphia's naturalists were particularly curious about the man. They pretended that their curiosity was "never prurient and aimless," one explained, and that disciplined curiosity gave that city its special character. But they weren't so very different from the ticket-buying Bostonians. In Boston Brooks learned that American Christians wanted to probe his faith, but in Philadelphia men wanted to measure his body and finger his skull. Philadelphia's students of nature reported that he measured five feet, appeared "thick set and strongly made," and had a "much distorted" skull.[3] Measuring men was the kind of thing these naturalists did, as their generation's contribution to the city's intellectual legacy. They carried on with a Quaker tradition that steered talented young men into medicine and science, rather than the ministry. Historians credit Philadelphia's nineteenth-century naturalists with helping to create America's scientific establishment in the amateur societies that flourished in the 1830s and 1840s. Naturalists studied plants and animals; discussed geology, chemistry, and astronomy; and developed a "science" of racial difference from work on human anatomy. They would not yet have called themselves scientists, but they kept up eighteenth-century ideas about observation and inquiry and celebrated discoveries that they thought broke with common conventions.[4]

The men who met that December at Dr. Blanding's house included craniologist Samuel George Morton, just then finishing his great anatomical atlas, *Crania Americana; or, A comparative view of the skulls of various aboriginal nations of North and South America*; ornithologist John Kirk Townsend, recently returned from a trek across North America; and Col. Thomas L. McKenney, a retired Indian agent and collector of Indian portraits. The well-dressed white men hustled Brooks into a stuffy drawing room. They had all met Native Americans on other occasions, but this meeting bristled with questions about human skulls generated by phrenologists and craniologists. In the 1830s, many educated men and women in Europe and America believed that the skull's external contours revealed inner character. A man's potential could be read in the shape of his head. In the eyes of these white Philadelphians, William Brooks had a curiously shaped head.

Each of these men needed something from young William Brooks. Lee hoped he could help prove to Methodist elders that despite the terrible epidemics that were killing native peoples in Oregon faster than they could be brought to Christ, there were still souls enough to make efforts at conversion worth the cost. Displays of native piety could also help Lee put rumors to rest that he was out in Oregon as part of the first wave of American settlement aimed to wrest the territory, once and for all, from English control. Diseases had reduced Oregon's native population and Lee's project did promote secular goals, but William Brooks and the story of his conversion gave Lee the chance to quiet those who questioned his efforts in the recently devastated, still thinly settled Willamette Valley.[5]

Phrenologist Combe needed Brooks too. Students of phrenology pressed for an answer to a "curious and important question." As one writer asked, what is "the mental effect of an artificial or accidental change in the shape of the skull?"[6] Combe also believed that he could support his case for phrenology if he could prove that people who flattened the skulls of infants were actually instinctive phrenologists, manipulating baby heads to cultivate faculties they wanted to see develop in adults. Had "Flatheads" shaped skulls to develop "Secretiveness" but to reduce "Benevolence"? Combe would never know for sure until he got hold of a flattened skull, preferably one with a fresh brain.

Craniologist Morton also had something to learn from Brooks's skull. When Morton met the young man in December 1838, he had been collecting human and animal skulls for almost a decade. But skulls from tribes west of the Mississippi were especially rare. As he wrote in thanks for a donation, "The skulls in question are particularly interesting to me, inasmuch as I have had great difficulty in obtaining any of those tribes." Intentionally deformed skulls were rarer still. And they commanded premium prices on the skull market. Collectors bargained hard for skulls reshaped by human hands and sometimes refused to lend them to fellow scholars. Morton sent out materials promoting *Crania Americana*, promising the book's subscribers descriptions of the "extraordinary distortions of the skull, caused by mechanical contrivances." "In fact, the author's materials in this department are probably more complete than those in the possession of any other person; and will enable him to satisfy the reader on a point that has long been a subject of doubt and controversy."[7]

Points of doubt and controversy were many, and for a craniologist the stakes in settling these questions were high. How malleable were infant skulls? How much could a mother shape a child's head? If skull shape could be created by men (or women), could craniologists count on skulls as natural markers of difference? Craniology's critics argued that the "characteristic diversities" of skulls were the "mere result of artificial causes originating in long perpetuated customs and nursery usages." In the middle years of the nineteenth century, Canadian ethnographer Daniel Wilson began to speculate that culture, not nature, determined a skull's shape. Women who carried babies in cradle boards or preferred to nurse on the left or right breast shaped their children's heads. According to Wilson, the only people whose skulls maintained a natural shape were ancient Egyptians and modern Fijians, who slept from childhood with heads carefully supported by neck pillows.[8] Malleable skulls threatened to undermine one of craniology's ruling premises that skull shape recorded racial differences. Morton needed to find the traces of the natural shape preserved in William Brooks's distorted skull. Since Brooks was alive, he could not explore his questions about the internal capacity of his skull.

And Brooks represented an opportunity for portrait publisher McKenney. Most of the Native Americans who appeared in McKenney's books were members of eastern tribes, men McKenney had met while serving as superintendent of Indian Trade and of Indian Affairs in the administrations of James Monroe and John Quincy Adams. McKenney and his colleague Judge James Hall began to put together their Indian portrait gallery in the 1820s. Many of their subjects were native leaders who posed during diplomatic missions to Washington, D.C. Despite clear evidence that there had been contact between these native leaders and Europeans and Americans, the portrait gallery preserved traces of traditional patterns. When Andrew Jackson fired McKenney in 1830, he moved to Philadelphia and continued to work on his portrait project. By 1837 he had signed up 1,250 subscribers, names enough to raise about $150,000. But McKenney's fortunes, along with those of his printers and many of his subscribers, were upended by the severe economic depression that hit the country in 1837. By the time the project limped to completion in 1844, it had run through five different publishers and cut deeply into the fortunes of its authors.[9]

With Brooks, he finally had a man from beyond the Rockies. But what did William Brooks make of his encounter with these skull men? Are there lessons to be discovered in the short life of a native man whose deformed skull fascinated American naturalists? Methodists, phrenologists, and naturalists have left records of their meetings with the unusual young man, preserving traces of Brooks's observations on the United States in the late 1830s.[10] We catch a glimpse of the complexly tangled history of American ideas. To capture Brooks's story and the history of a little fracas over his skull, we need to weave together chapters in the story of nineteenth-century America that we often hold as separate experiences of native families along the Columbia River, of Methodist missionaries holding an outpost in Oregon, of craniologists in Philadelphia hatching unfortunate arguments about superior and inferior races, of phrenologists feeling the bumps on American heads.

"Wm Brooks, a Native Flathead Indian"

After meeting William Brooks at Dr. Blanding's house, Col. Thomas L. McKenney invited him to sit for a portrait. The likeness of Brooks appears in the second of three volumes of McKenney's *History of the Indian Tribes of North America*. McKenney called him "Stum-A-Nu. A Flat Head Boy," preferring that identity in a book of Indian portraits to "William Brooks, a Christian Man." He posed Brooks in an eye-catching yellow-and-black blanket. The picture captures us with its beauty, but the title and the blanket hint at the image's deceptions.

In McKenney's collection of chiefs and warriors, Brooks stands out as a young man, distinguished by his flattened head. McKenney or his artist (scholars aren't sure who painted the picture) did what they could to dress him up and draped the youth in a striking Chilkat blanket, which had been made by the Tlingit people of southeastern Alaska,

Fig. 5. "Stum-A-Nu. A Flat Head Boy," from Thomas Loraine McKenney, *History of the Indian Tribes of North America* (1836–44).
(Courtesy of the Smithsonian Institution Libraries, Washington, D.C.)

not by Brooks's Chinook relatives. Tribesmen, European traders, and American collectors prized these blankets, tracing them to prominent families and eminent chiefs. Lee may have traded for the blanket in Oregon and carried it east with him, recognizing that it made a good prop for his fund-raising sermons. In fact, one Methodist congregant remembered Lee exhibiting "a curiously wrought blanket" that "was spun and wove in an ingenious manner, very thick and covered with hieroglyphs." At home, William Brooks would not have worn this thick and beautiful blanket.[11]

While Brooks posed, McKenney asked him about his life. Brooks spoke English well enough for Morton to compliment his accent and grammar, but Lee probably helped craft the life story McKenney wrote. McKenney's chapter on "A Flat Head Boy" begins by addressing the confusion about the names Americans had given the native peoples of the far West: the "honest, hospitable, and kindly disposed" people Americans called the Flatheads did not, in fact, flatten their heads. The "savage custom" of head flattening characterized the Chinook, an "ignorant and timid" people who lived farther west. McKenney repeated reports of the marvelous exploits of Chinook people in oceangoing canoes, acknowledged that trappers and traders had brought them into American and British orbits, and then turned to Stumanu's life.

The young man had been born about 1818, seven miles from the mouth of the Columbia River. He told McKenney that his father had died when he was two and an uncle raised him, his younger brother, and his sister. The uncle taught him to steer a canoe and to fish. Even though his people had been dealing with Europeans and Americans for a generation or more, Brooks recalled for McKenney a world where the patterns of traditional life—babies strapped to cradle boards on their mother's backs, men and women fishing for smelt and salmon and trading the surplus for horses with people who lived farther inland—still prevailed. (Gender equality impressed observers, who visited these people whose principal occupations were not hunting and war.) In keeping with his smiling portrait, Brooks spun a tale of Chinook life in the 1820s, touched by white trappers and traders but not yet ravaged by rum or by the epidemic diseases that forever altered his world.[12]

McKenney preserves few details, but evidence suggests that the boy's life changed when his uncle died in an epidemic of "intermittent fever" that devastated the villages along the Columbia River in the early 1830s. Demographers believe that the disease (probably a strain of malaria particularly deadly to the native peoples) cut the population of Chinookian and Calapuyan people by a staggering 88 percent. Lewis and Clark estimated the population at around 15,000 in 1805. By 1841 the tribes numbered 2,000 people. Biological destruction pulled the culture down with it. The orphaned children sought shelter with whites, turning first to Dr. John McLoughlin, chief factor of the Hudson's Bay Company, the most powerful white man in the region and an unhappy witness to the fever's "dreadful havoc" on native families.[13]

McLoughlin sent the children on to Jason Lee's recently established Methodist mission on the Willamette River. McLoughlin was pleased to have Lee's help taming the Oregon Territory. Lee's principal congregants were Hudson's Bay Company trappers who had settled in the Willamette Valley and begun to raise mixed-race families with native women. Lee and his colleagues built a school and clinic. In McKenney's telling, Stumanu took "his younger brother by the hand, proceeded to the school, to offer himself and his brother as pupils." A younger sister came along too. Missionaries counted these children ("poor savages of the farther west," McKenney called them) particularly valuable members of a struggling mission settlement, short on both farmhands and Indian souls. "The special providence of God has, already, seemed to throw upon our care three poor Flathead orphans; one, a lad of fourteen or fifteen years of age, who is quite serviceable in several ways," the mission teacher reported to eastern readers in 1835. "These children came to us almost naked, in a very filthy state, and covered with vermin. . . . *J. Lee* cleansed them from their vermin, so that they do not now appear like the same children they were when they first came."[14]

Lee assigned the children tasks to help assure their material and spiritual well-being, and Stumanu "quickly showed a great fondness, as well as an aptitude, for learning, was industrious and useful on the farm, and won esteem by the most amiable qualities of temper." According to McKenney, he also possessed "what was remarkable in an Indian, a decidedly mechanical genius, and excelled in the construction of tools and implements, and in the imitation of any simple articles of furniture that came under his notice, so that the mission family were fully repaid for the expense of his education and subsistence by his labour." Stumanu got food and clothes in exchange for his labor and earned a new name when Jason Lee baptized him "William Brooks" in memory of a Massachusetts clergyman. In the same way, Stumanu's sister Kye-a-tah became "Lucy Hedding." These given names made the children part of the "Christian family," although they helped to hide their family ties and their ancestral associations as well.[15]

It was remarkable that the children survived and stayed healthy enough to contribute to the mission enterprise. Lee and his colleagues kept a sort of "grim ledger," listing names of children who died or ran away. Those who died before they had worked long enough to cover the costs of a simple pine coffin appeared as deficits in mission account books. Fifty-two pupils showed up during the school's first four years. At least eight died; most others ran away or were taken away by their parents. Stumanu, his brother, and his sister had no family and nowhere else to go.[16]

The mission children lived between worlds, straddling identities that must have pulled hard in different directions. The missionaries baptized the children with their new names. They banned their native language and taught them to speak English and to read and write. The children sang Methodist hymns and plowed and planted, like good American farm children. But the missionaries admitted there were advantages in having the children retain some of their native skills. Lee set out to produce a generation of native farmers, but one of his colleagues was pleased by a boy who spent "the forenoon in hunting (by which we are supplied with some animal food every day) and the afternoon in learning to read." Closer to Lee's project, to raise money in the East, he needed "William Brooks" still to appear to be "Stumanu," both Christian convert and flat-headed "savage."

William Brooks lived and worked at the mission school for about three years. Hopes that he could pull his family into a Christian future must have been destroyed when his sister died of scrofula (a vague description of any one of several diseases, including tuberculosis, that infected the lymph glands) in October 1837 when she was about fourteen. "We took her to Vancouver for the benefit of medical advice," one missionary wrote. "She tarried a few weeks, but was discontented and homesick. She arrived on Friday and died on the following Thursday."[17]

After his sister's death, William's life changed again when the Reverend Lee took him along on the fund-raising tour in the East, which brought them to Philadelphia. Lee described the trip as a "Duty," which "required me to leave home and wife and friends and retrace my steps to the land of civilization." He planned to recruit "a mission steward or business agent, two carpenters, a cabinetmaker, a blacksmith, and two farmers." The list seems to confirm Methodist fears that Lee's mission had taken a secular turn. Contemporaries recognized that it was difficult to keep spiritual ends separate from secular means. The mission settlement—with its manual labor school, a store, and a hospital—did serve as an arm of American colonization in Oregon, and Lee, "a tall and powerful man, who looks as though he were well calculated to buffet difficulties in a wild country," had talents suited for work in this world.[18]

Little wonder that Lee found it challenging to balance the earthly needs of settling the country with the spiritual expectations of a Methodist mission board in New York. The future must have seemed very cloudy. As Lee watched the native population decline, he wondered if intermarriage could keep native blood running in the veins of all Oregonians. Backers needed more tangible evidence of the mission's effectiveness in bringing native souls into the Christian fold, so in the spring of 1838, Lee set out for the East with "two Indian boys—Wm. Brooks (a Chinook) and Thomas Adams." Lee also took along three mixed-race sons of a Hudson's Bay Company employee, William McKay, and promised to drop them at a boarding school in Massachusetts.[19]

A first plan to reach New York by traveling west across the Pacific and then around the world on a Hudson's Bay Company ship fell through, and Lee and his young charges set out overland from Fort Vancouver in April 1838. Lee left his wife behind, a woman he had married eight months earlier. Methodist elders sent Anna Maria Pittman west, expecting Lee to marry her. With candor, Lee admitted he did not fancy Anna at first, but "at length became convinced that she was eminently qualified to do all the duties and kind offices of an affectionate companion." Duties and kind offices left Anna six months pregnant when Lee headed east over the Cascades.[20]

Lee and his charges spent a night at Marcus and Narcissa Whitman's Presbyterian mission in the Walla Walla Valley, where Lee preached in the Chinook jargon to a congregation of "Indians and Kanakas," that is, people native to the Columbia region and native Hawai'ians employed by the Hudson's Bay Company. That difficult linguistic exercise perhaps increased his sympathy for William Brooks's later efforts to preach in English to American congregations. Lee hoped to leave the Whitmans' establishment in eastern Oregon in time to join up with fur trappers heading for the summer rendezvous—the annual sessions where white and native trappers met the pack trains from St. Louis, traded their beaver pelts for money and goods, told stories, and drank. Lee and his group missed the meeting and the chance to travel east with St. Louis traders, but he gathered a few traveling hunters. The party headed east, moving against the current of America's westbound traffic. Lee insisted the group rest on the Sabbath, when he tried to preach God's word to his "small, sleepy and apparently indifferent congregation."[21]

Sad news that Anna and her baby had died in childbirth reached Lee in July when the party stopped at a Methodist mission among the Shawnee on the Missouri frontier. Pioneer chroniclers honor Anna as the "first white woman to die in Oregon." Lee mourned for his wife and newborn son, but a journal he kept during those weeks hints at the trip's pleasures. Lee boasted that he and Brooks each brought down a buffalo. Lee killed his with his third shot. "Wm. also killed one," he wrote. "We thought we did very well, as there were but seven buffaloe, and so many old hunters, considering this was our first trial."[22]

Lee soldiered on east and dropped McKay's boys at their appointed schools (two of them at his alma mater Wilbraham Academy in Massachusetts). Thomas Adams took ill on the overland journey and spent several months in Peoria, Illinois, before rejoining Lee and William Brooks on their eastern tour. Brooks and Adams helped Lee raise money—the three of them representing Methodist success in the mixed-race mission out west. Large crowds turned out as Lee sermonized up and down the eastern seaboard in the winter of 1838–39. Their presence began to answer a question that nagged some eastern supporters: "What have they been doing all this time in Oregon?" The answer came as part of a package that mixed pious instruction with entertainment.

Again, Lee did not pioneer the Native American performance, either on stages or in churches, and many in the Methodist congregations had a chance to have seen George Catlin's traveling exhibitions of Indians, artifacts, and paintings. (In fact, during the early spring of 1839 before Catlin sailed for London with his native performers and his "Indian Gallery," curious easterners could have gone to see Catlin's group one night and to listen to Lee and Brooks the next.) The religious convictions, far western origins, and unusual appearance of the two young men drew large crowds. Lee learned to make money for his settlement at just those points where the sermon shaded into a show of curiosities, where piety mixed with popular entertainment. Congregants liked seeing and hearing Adams and Brooks.[23]

On a December visit to Washington, Brooks delivered his first public speech in English. Lee expected "a failure" but "was agreeably disappointed." Two Ohio congressmen assured Lee "that they were greatly and agreeably surprised . . . that they could not have anticipated such shrewd, appropriate, and intelligent remarks from a youth under such circumstances." Over the next months, audiences responded well, crying, laughing, and donating money, even though Lee competed for scarce dollars with advocates for Methodists missions in Africa. Brooks's "tears spoke with resistless eloquence" to a Hartford man, who remembered the young man's tribute to the late Mrs. Lee. Sympathetic

Massachusetts shoe workers gave him a new pair of boots. His candor sometimes "excited the risibles," as Lee put it, but laughing audiences helped the mission's finances. Lee kept a tally of the money he raised, thrilled when figures ran above $500. He had great success with the Methodist Board, which approved an allocation of $40,000 for the Oregon mission. (That figure comes close to a healthy $1 million in today's money.) Supporters also donated seeds, farm machinery, and parts for a gristmill and a sawmill. Brooks's eloquence at missionary meetings made this possible.[24]

The young man pleased other potential supporters with his good table manners ("yet I never in a single instance saw him, by accident or through ignorance, do anything that would be considered *outlandish* even by the polite or well-bred," Lee wrote), apparent piety, and striking speeches. Lee recalled, "Seldom did he arise to address a congregation without bringing forward something new and striking that he had not mentioned in any previous address; so that, contrary to what might be expected, his daily communications, instead of becoming stale and tiresome to me, by their tame monotony, were always interesting, and sometimes delightful, pathetic, and thrilling, even beyond anything I had dared to hope from him."[25]

Accounts describe Brooks's skill with his audiences and suggest that he liked expressing his opinions, although Methodist mediation makes it difficult to be certain we recover Brooks within his speeches. We can catch him needling his hosts for hypocrisy and adding his observations on the people and things that impressed or amused him on his tour of the East. The Methodists worked the young man hard, organizing addresses throughout northern New England in the bitter cold of a January and February. Lee remembered Brooks drawing large audiences, night after night. The two men spoke, but then stayed on to answer "innumerable questions." Sleep gave Brooks some relief, but it was "a small portion of the time, for the good friends where we lodged seemed often to forget that we needed rest, or at least they seldom thought of it till twelve or one o'clock, even when we had to start at four or five in the morning."[26]

Over the months, Brooks grew more confident and began to criticize the hypocrisy of Americans he met. Early in 1839, three or four months after Lee and Brooks had arrived in the East, Methodist journals reported that the young man now wondered "at the wickedness he saw." In his "somewhat ambiguous English," he chastised those who saw the mission in Oregon as only the beachhead of a secular settlement. Brooks was proof that good Christian work was going on out in Oregon, and his presence helped quiet some of Lee's critics, who suspected him of just such worldly ambition. Brooks also made observations of his own. He was particularly struck by a blind man he met in Baltimore. "He's [a] colored man—he belongs to our Church. He can't read, he can't see nothing, but he sees Jesus Christ. Children, you say that old blind man, colored man, miserable—but he be very happy. O, I love that old man, because he love Jesus Christ."[27]

Brooks also complained that white rum sellers carried death over the Rockies. "White men settled there about twenty-five years ago. Indians great many years ago never died very fast. But since settlement [by] white men, they have died every hour. They give them rum—everything that is bad. I tell you one thing I want you to put down on paper, that *I don't want any wild Yankee there.*" And he didn't much like the "rude and wicked" children who insulted him, staring and laughing as he walked by. With some saucy questioners, however, he apparently joked about his flattened head. Several Methodist chroniclers recorded the story of a woman who asked Brooks about his skull, "rallying him somewhat on the curiosity of the fashion. William replied, 'All people have his fashion. The Chinese make little his foot. Indian make flat his head. You,' looking at her waist and putting his hand to his, 'make small here.' She at once decided if his head was flat, his wit was sharp." Or so the story goes.[28]

We can follow this anecdote back to Brooks's visit to Philadelphia in the winter of 1838 and to the new things that Brooks might have encountered there or heard about in conversations with his new acquaintances. As he walked the streets, Brooks noticed women's waists cinched by whalebone corsets, but there was only one place he would have seen the bound feet of a wealthy Chinese woman, even if he had run into Chinese sailors from time to time back in Oregon. On Christmas Eve 1838, while Lee and Brooks were still in the city, China merchant Nathan Dunn opened his collection of "Ten Thousand Chinese Things" on the ground floor of Peale's Philadelphia Museum. The show was the season's sensation, drawing some 8,000 visitors during its first week, and somewhere close to 100,000 before it closed in 1842. Visitors walked through a lacquered gate and into "China in miniature," as one put it. There were dozens of dioramas depicting scenes of everyday Chinese life. Merchants, mandarins, literati, beggars, boat women, barbers, blacksmiths, jugglers, and shoemakers, all performing their ordinary tasks. And then there were three hundred cases filled with Chinese things: lacquered boxes, printed books, porcelain teacups, cushions, vases, hats, candles, shells, birds, fish, and a thirteen-foot boa constrictor. Hundreds of paintings—portraits and landscapes—lined the walls.

"In short," phrenologist Combe concluded after touring the collection, "a survey of this museum approaches closely to a visit to China." A museum visitor "is present at the interviews of its mandarins; he sits with the ladies in their morning parties, he is admitted into the studies of literati, and examines their books; he sees face to face, the priests of different sects; he buys his goods in the trader's shop; he is jostled by the sedan and bearers; . . . he passes by the barber, the shoemaker, and the smith, the soldier, and the boatwomen, the servant and the beggar; . . . he goes to the theater and sees the actors at night; and he passes a cool and tranquil day in a visit to the country summer-house of a rich merchant." There were also lessons on agriculture and manufacture.[29]

For many Philadelphians, a visit to the Chinese display was a chance to get a sense of the world behind the China trade that was making a handful of neighbors so very rich, some of them by trafficking opium from Turkey. (These were dedicated American free traders determined to cut into the British monopoly. It was all legitimate commerce, they argued, even if it was commerce in a poisonous substance banned by the Chinese government.) Dunn made his fortune in China, without touching the trade in narcotics he thought illicit, and he had developed a deep appreciation for the beauty of China's material culture and the exquisite skill of Chinese manufacture. He built a Chinese country house outside Philadelphia. When the sensation around the Philadelphia show faded as sensations do, Dunn packed the collection off to London, where England's Opium Wars sparked new interest from that city's crowds.[30]

Philadelphians went to the show to satisfy their curiosity about the world and its people. Like the naturalists who came out to meet William Brooks, many Americans during the 1830s and 1840s were trying to figure just who they were by exploring who they were not. Comparison could take the form of vicious racism. But sometimes curiosity and comparison spawned a gentler cultural relativism. By needling his audiences about their cinched waists, Brooks tried to tease Americans in this direction.

A visit to Dunn's "Chinese Things" might have given him material for his object lesson. He and Lee could have gone along with the first week's excited crowds. One exhibit particularly interested visitors who had been thinking about manipulated bodies. Dunn displayed "the bandaging of the feet"; "golden lilies," he said they were called. Bound feet marked rank in China, "just as small and white hands are with us deemed proof of gentility." "Civilized" Philadelphians shuddered, but Dunn did not judge. "This is, no doubt, an absurd, cruel and wicked practice; but those who dwell in glass houses should not throw stones." "All people have his fashion," as Brooks put it. Bound feet actually damaged women's health less than tight stays, Dunn wrote. Brooks knew that his distorted skull did not damage his health or that of his kinsmen. In fact, like bound feet, a flattened head served to distinguish him as superior to the round-headed captives, servants, or slaves living among his flat-headed people.[31]

"Abstracting the Skull"

In December 1838, the odyssey Brooks had begun among the Oregon Methodists crossed into the territory of the Philadelphia naturalists. Brooks's wit seemed sharp to Morton, too, and he thought Brooks had "more mental acuteness than any Indian I had seen." The craniologist also reported that Brooks was "communicative, cheerful, and well mannered." To look at Brooks, Morton said, one noticed his "marked Indian features, a broad face, high cheek bones, large mouth, tumid lips, a large nose, depressed at the nostrils, considerable width between the eyes, which, however, were not obliquely placed, a short stature and a robust person."

Brooks "cheerfully consented" to let the good doctor measure his head. Morton recorded a longitudinal diameter of 7.5 inches, parietal diameter of 6.9 inches, frontal diameter of 6.1 inches, breadth between the cheekbones of 6.1 inches, and a facial angle of about 7.3 degrees, although he could not guess at the skull's capacity.[32] Other craniologists did not find their subjects so cooperative. "It is no easy thing to obtain actual measurements of Indians' heads," wrote Daniel Wilson, the Scottish historian, ethnographer, and craniologist working in his adopted

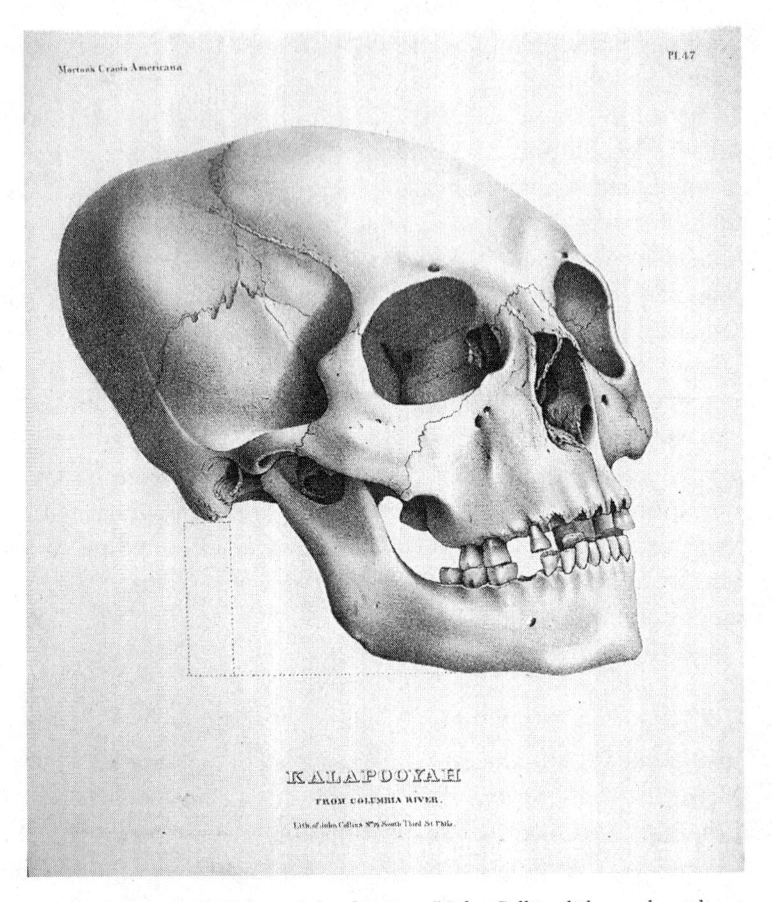

Fig. 6. "Kalapooyah Skull from Columbia River," John Collins, lithographer, plate 47, from Samuel George Morton, *Crania Americana* (1839).

Canada. "I have seen an Indian not only resist every attempt that could be ventured on, backed by arguments of the most practical kind; but on solicitation being pressed too urgently, he trembled, and manifested the strongest signs of fear, not unaccompanied with anger, such as made a retreat prudent."[33]

If this patient man knew that Morton coveted his skull, if he thought for a moment that he had wandered unwittingly into a land of headhunters, he did not let on. But Morton let slip that "what most delighted me in this young man, was the fact that his head was as much distorted by mechanical compression as any skull of his tribe in my possession, and presented the very counterpart" to the "Kalapooyah" skull he pictured in his book. ("Kalapooyah" was Morton's preferred spelling. Others used Calapuya.) Morton had purchased the skull he imagined as Brooks's counterpart from John Kirk Townsend, the ornithologist who joined the group at Dr. Blanding's house. Morton reported that Townsend "knew this young man . . . in his own country, and Townsend and Brooks recognized each other when they met in Philadelphia."[34]

Townsend was born in Pennsylvania in 1809. His Quaker relatives were doctors, dentists, and naturalists. John Townsend studied to be a dentist, like his brothers, but birds were his real love. Townsend was a good ornithologist, and he worked hard. But John James Audubon was better. Audubon's success gnawed at Townsend. He complained that Audubon never acknowledged the ninety-three specimens he caught and stuffed for him, and he wondered how he could produce a book of American birds to rival Audubon's great project. After a brief stint as a taxidermist at Washington's National Institute for the Promotion of Science (a Smithsonian precursor), Townsend went back to taking care of teeth.

In the end, his bird work did him in. Townsend died in 1851 at the age of forty-two poisoned by the arsenic in the powder he had developed to preserve dead birds. Curators regretted the death of this talented taxidermist, a man whose specimens were mounted "in the most splendid manner." "Townsend can skin, stuff and sew up a bird, so as to make it look far superior to any I have seen, in five minutes," museum director Spencer Fullerton Baird remembered. Museum workers knew the hazards of the work: "The immense quantity of arsenic and corrosive sublimate necessary for their preservation requires, respectively, that very great caution should be observed, and that the handling and arrangements should be under either the immediate inspection or personal attention of one fully adequate to all the details connected with this subject. *In the hands of inexperienced persons, death might be the result.*"[35] And so it was for Townsend, an experienced man, but a quick-working and careless one.

When he went adventuring in the American West in 1834, Townsend was young, enthusiastic, and not yet spoiled by competition among America's naturalists. He took a steamboat from Pittsburgh out to St. Louis, where he joined an expedition organized by Boston ice-merchant Nathaniel J. Wyeth, the man who made a fortune when he discovered that New England ice, packed in sawdust, could be shipped around the world. (Wyeth, who planned to make a second fortune in Oregon's fur trade, joined Lee in the roster of America's Oregon entrepreneurs, imagining an American future in territory the country had yet to claim.) Townsend made his way west on foot, walking slowly, looking at the landscape, and hoping to chance on a species of bird or rodent that could surprise city-bound naturalists. The 1830s left only a few opportunities for a man to christen a new species, preserving his name in the annals of science. But Townsend gobbled up what he found on that trip, discovering a shrew mole, a meadow mouse, a marmot, a ground squirrel, a great-eared bat, a hare, a gopher, a thrush, a sandpiper, a cormorant, and a warbler that he named after himself. He could be generous too. He "honored," as he put it, a species of water ouzel (the *Cinclus mortoni*) "with the name of my excellent friend Doctor Samuel George Morton, of Philadelphia."[36]

Once in Oregon, the naturalist turned his attention to skulls. Before leaving Philadelphia, he had promised his friend Dr. Morton that he would try to secure a flattened head or two for his skull collection. He wrote from the West that he had risked his life to rob an Indian burial place in Oregon, sneaking off with a reeking pack of human remains. He told Morton it was "rather a perilous business to procure indian sculls [*sic*] in this country. The natives are so jealous of you that they watch you very closely while you are wandering near their mausoleums & instant & sanguinary vengeance would fall upon the luckless night [*sic*] who should presume to interfere with their sacred relics." "Great secrecy is observed in all their burial ceremonies, partly from fear of Europeans," an English traveler commented. Lt. Charles Wilkes, who brought the ships of the United States Exploring Expedition to the Oregon coast in 1841, warned that "to rob their burying grounds of bodies, is attended with much danger, as they would not hesitate to kill any one who was discovered in the act of carrying off a skull or bones."[37]

Townsend also described the techniques tribal mothers used to flatten infant skulls. He found the people of the Columbia River region "shrewd and observant," as did most European and American visitors, and concluded that skull flattening did not harm the intellect. But the process seemed to pain infants. A mother placed her baby in a "sort of cradle . . . formed by excavating a pine log to the depth of eight or ten inches." The cradle was lined with a bed of grass mats, but a child was bound by a "little boss of tightly plaited and woven grass" fastened over the forehead and tied on the cradle's sides. "The infant is thus suffered to remain from four to eight months, or until the sutures of the skull have in some measure united, and the bone become solid and firm. It is seldom or never taken from the cradle, except in case of severe illness, until the flattening process is completed."[38]

Townsend wrote that a skull properly flattened remained the mark of a child well raised. With that thought in mind, he may have noted that William Brooks's flattened head registered his mother's plans for the boy's good future among his own people. Brooks's flattened head leaves evidence that she believed in the rightness of her people's customs, culture, and rituals. Even as the world along the Columbia was changing, she flattened her babies' heads to make them her own, to make them members of her tribe. In this world, grown-up tribal leaders had flattened skulls; captives and slaves—individuals taken from other tribes—did not, their unflattened heads marking them permanently as outsiders, inferiors in these strongly hierarchical societies. As one mid-century observer noted, "The Flathead tribes are in the constant habit of making slaves of the Roundheaded Indians; but no slave is allowed to flatten or otherwise modify the form of her child's head, that being the badge of Flathead aristocracy."[39]

Yet Townsend's language also suggests that he found skull flatteners somewhat less than human. He admitted he was frightened and disgusted by a young child recently removed from a cradle board—a child that remains an "it" in his account: "Although I felt a kind of chill creep over me from the contemplation of such dire deformity, yet there was something so stark-staring, and absolutely queer in the physiognomy, that I could not repress a smile; and when the mother amused the little object and made it laugh, it looked so irresistibly, so terribly ludicrous, that I and those who were with me, burst into a simultaneous roar, which frightened it and made it cry, in which predicament it looked much less horrible than it had before."[40]

Townsend witnessed the beginning of a life with a flattened skull, but skull collectors really began their work when lives ended. Flattened heads brought a premium in the skull market. Members of western tribes tried to stop European and American collectors who snatched bodies from burial sites. The epidemic diseases that destroyed so much of the population in the 1830s simplified the work of collectors, but that history brings tragedy to Townsend's adventures. "I have succeeded in hooking one," he told Morton when he got his hands on a skull, "& no doubt in the course of the winter, I shall get more. There is an epidemic raging among them which carries them off so fast that the cemeteries

Fig. 7. "Chinook Woman and Child," *Harper's Magazine* (1870). (Picture Collection, The New York Public Library, Astor, Lenox and Tilden Foundations.)

will soon lack watches. I don't rejoice in the prospect of the death of the poor creatures certainly, but then you know it will be very convenient for my purposes."[41]

Casual language of convenience describes a people so devastated by disease that the living could no longer bury the dead, let alone keep watch over corpses and newly buried bodies. Some coastal tribes buried bodies in canoes or scaffolds. Had Townsend been on the Pacific coast in the 1820s, he might have watched Chinook survivors dig a grave for the body of the newly dead and bury a corpse along with all the clothes that had belonged to the once living individual. Survivors arranged other possessions—"wooden spoons, hats, tin kettles, beads, gun barrels bent double, and tin pots"—on carved headboards. Like dead bodies, possessions were to stay where they were placed. "Although they are very superstitious about disturbing the articles belonging to the dead, yet these all have holes punched in them, to prevent their being of any use to others, or a temptation to their being taken off," a visitor wrote in the early 1840s.[42]

As burial practices crumbled under the weight of mass death, Townsend saw people pushed to their limits. Physical markers of native culture survived for a time in Stumanu's flattened head, but the lice-covered children were left orphans in a devastated world. Depopulation along the Columbia "has been truly fearful," he wrote. Witnesses to the effects of "intermittent fever" told him of devastated villages. One man remembered counting "no less than sixteen dead, men and women, lying unburied and festering in the sun in front of their habitations." And he added a horrifying description: "Within the houses all were sick; not one had escaped the contagion; upwards of a hundred individuals, men, women, and children, were writhing in agony on the floors of the houses, with no one to render them any assistance. Some were in the dying struggle, and clenching with convulsive grasp of death their disease-worn companion, shrieked and howled in the last sharp agony." The scenes Townsend described lived in Brooks's mind as childhood memories.[43]

Escalating death rates strained traditional ways, making it difficult for people, even those skilled at adapting to new ways, to follow customary mourning and burial practices. While desecrating graves, Townsend acknowledged that proper disposition of the dead helped to anchor a people in place and in time. He knew that conquest (and conversion to Christianity) transformed these spatial and temporal connections. Skull hunters had their own small parts to play in this drama of destruction. Brooks's reshaped head may have been an expression of his mother's faith in the future; Brooks's proper care for his mother's dead body would have helped carry his people's past into that future. But events in Brooks's life in the 1830s troubled Chinook connections with the past as surely as they altered the Chinook roads to the future his mother had imagined for him.

Townsend headed home to Philadelphia, taking a long trip by sea to Hawaii and on around the world. He stopped long enough in Oahu to pick up the skulls of a ten-year-old girl and a forty-year-old man, which he brought to Morton. But he turned his time in Oregon into a *Narrative of a Journey across the Rocky Mountains to the Columbia River and a Visit to the Sandwich Islands*, which appeared in 1839, shortly after he returned to the States. Skull collecting spiced up his story. He acknowledged that grave robbing could be "sacrilege," as he put it, and he accepted disease as a skull collector's ally, but then he spun his exploits into tales, embellished with details that give them the air of romantic adventures.

Here is an example. "During a visit to Fort William," he wrote, "I saw . . . a canoe, deposited, as is usual, in the branches of a tree, some fourteen feet from the ground. Knowing that it contained the body of an Indian, I ascended to it for the purpose of abstracting the skull." Instead of a skull, Townsend found "a perfect, embalmed body of a young female, in a state of preservation equal to any which I had seen from the catacombs of Thebes"—the Egyptian measure of mummy making. He returned for the mummy that night, "at the witching hour of twelve," armed with a rope to help lower the body to the ground. He packed the lovely mummy into his canoe and paddled off. Back at the fort, he wrapped the body in a mat and told anyone who asked that he had a "bale of guns." Indian porters, who would have refused to move the body, happily packed his "guns" to Fort Vancouver, where he could ship things east.

In Townsend's telling, the fort commander got wind of the story when the brother of the poor dead woman complained that Townsend had disturbed the burial. He "had been in the habit of visiting the tomb of his sister every year. He had now come for that purpose . . . and his keen eye had detected the intrusion of a stranger on the spot hallowed to him by many successive pilgrimages. The canoe of his sister was tenantless, and he knew the spoiler to have been a white man, by the tracks upon the beach, which did not incline inward like those of an Indian."[44]

Good relations between natives and traders did not matter to Townsend, but the fort commander warned him that his collecting put his people at risk. He insisted Townsend return the body to the mourning brother and offer him "a present of several blankets, to prevent the circumstance from operating upon his mind to the prejudice of the white people. The grieving Indian took the body of his sister upon his shoulders, and as he walked away, grief got the better of his stoicism, and the sound of his weeping was heard long after he had entered the forest."[45]

Fancy tales aside, Townsend contributed to the collection of human

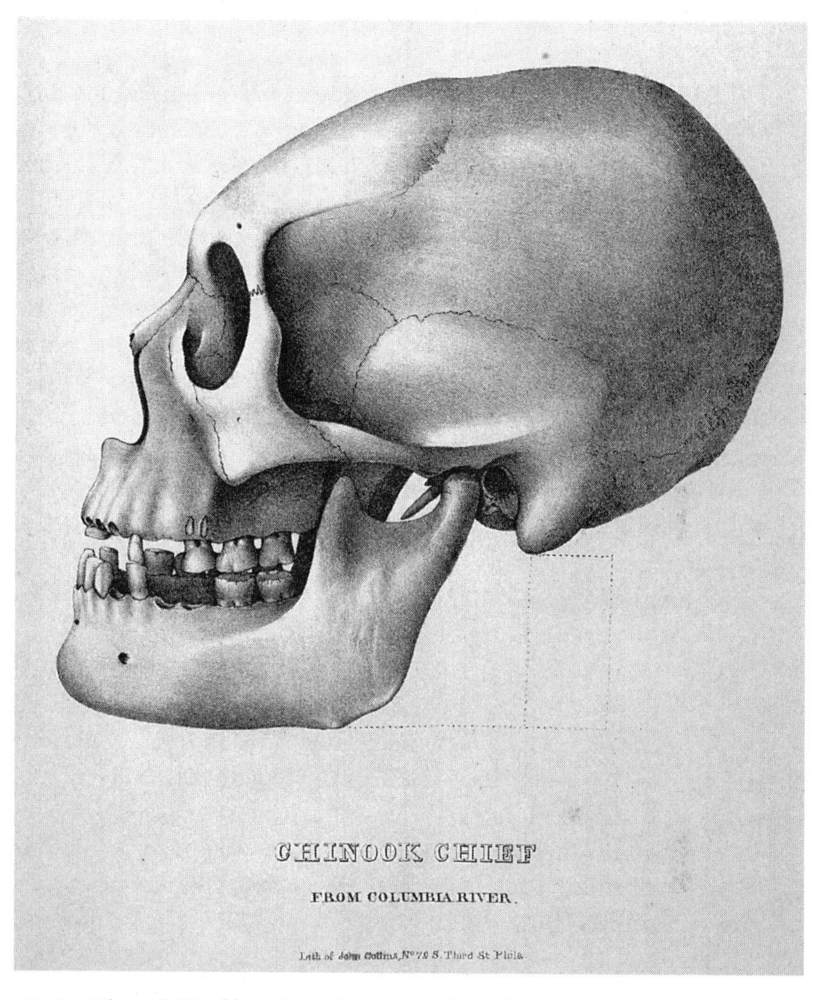

Fig. 8. "Chinook Chief from Columbia River," John Collins, lithographer, plate 43, from Samuel George Morton, *Crania Americana* (1839).

skulls that Morton was assembling in Philadelphia. Morton paid Townsend $50 for two flattened skulls, a high price that reflected their rarity and compensated Townsend for the dangers he said he had faced. In a final tally, Morton thanked Townsend for Hawaiian skulls, for the skull of the "Kalapooyah" flattened like William Brooks's head, for the heads of a chief and a child, and for a skull, not flattened, that Townsend assumed was the head of a captive or slave.[46]

"The Brains of These Flat-Headed Indians"

It would be interesting to know if Townsend recounted his Oregon exploits when he met William Brooks again in Philadelphia. Or if Morton showed Brooks the "Kalapooyah" skull he thought "his very counterpart." Even if they were discreet with young Brooks, skull stories simmered along just below the surface. A few months after that January meeting, Brooks's skull became the object of a sorry little squabble among the East Coast intelligentsia.

In the final paragraph of his biography of the Chinook man, McKenney reports, that "on the eve of the departure of the Rev. Mr. Lee to the scene of his labours on the Wallamette [*sic*], Stumanu, flushed with the prospect of once more mingling with his kindred and friends, and gratified with all he had seen of the white man's capacity and powers was taken suddenly ill, in New York, and after a short but severe attack, died on the 29th of May, 1839." The editors of the *New York Herald* noted among the dead "on Wednesday, 29th inst., Wm Brooks, a native Indian, aged about 20 years."[47]

Jason Lee diagnosed Brooks's last illness as a mix of pleurisy, "congestive fever," and typhus, but again tuberculosis killed him. (He could have picked up the infection in crowded mission dormitories, in the mission congregations that sometimes included sailors, or in the press of crowds in Philadelphia and New York. The winter trek through the icy damp of northern Vermont and the long nights answering questions from inquiring Methodists could not have boosted the resistance of a young man from the Oregon coast.) Methodist histories report that Lee watched over William Brooks in his last illness with "the care of a father." Shortly before he died, these histories report that Brooks told Lee, "'I want to go home.' 'What home?' said Mr. Lee; 'your home in Oregon?' 'No; my heavenly home.'"[48] Did such a conversation take place? It may be that William Brooks on his deathbed has been made to serve the purposes of a pious fable, selected for a part in the Sunday school books handed out to Methodist children in the 1840s and 1850s. Or maybe his thoughts mixed the images offered by McKenney and Lee. Maybe he died, as McKenney put it, "flushed with the prospect of once more mingling with his kindred and friends," but picturing that meeting in Lee's Christian heaven come alive with native dead. Or in an afterlife that mixed the stories of ghosts and guardian spirits he had heard as a child with tales he learned from Lee.[49]

Indeed, Lee's account of Brooks's Christianity is laced with images of death, suggesting that the new religion offered Brooks some real means to cope with the devastating experiences of his childhood in Oregon. Brooks clearly associated white presence with the escalating death rates among his people. "Indians great many years ago never died very fast. But since settlement [of] white men, they have died every hour," as he put it. Fortunately for Brooks, Christianity gave him new ways to think about the troubling rate of death. Christianity's promise of "everlasting life" caught his attention first. As Brooks explained, "Suppose one man believe in Jesus Christ and love Jesus Christ, by and by, when he die, he live with him in heaven; he never die a second death."

In many of his speeches, Brooks used the phrase "second death"—a means of describing the promise of salvation that missionaries took from the book of Revelation—"Over such the second death has no power, but they shall be priests of God and of Christ, and shall reign with Him a thousand years" (20:6). But the idea must have had a particular resonance for one who had seen as many "first deaths" as Brooks had seen. Conversion to Christianity seems to have provided Brooks with emotional tools to deal with death, just as it provided a metaphoric language to address eastern congregations. "I never go back again in the darkness heart. I pray more and more, and I go on. All time I thinking about those poor Indians dying in second death. I cannot sleep every night." He passed those sleepless nights contemplating a "representation of Christ upon the cross," which he hung in front of his bed. "I can't give it to you," he told one young woman who asked to have it as a souvenir of her encounter with the unusual youth. "I must carry it out among those poor Indians, and always show them Jesus Christ died for them, and then they will believe."[50]

The Philadelphia *North American* covered the New York funeral of this "estimable youth, beloved by all who were acquainted with him. But the best of all is, he died an experienced Christian." "The corpse was taken to the Greene-st. church on Thursday, and an address delivered on the occasion by Rev. Dr. Bangs." Brooks's body was buried in New York City near the Bedford Street M. E. Church. (And there it stayed until 1910, when the city of New York extended Seventh Avenue south of Bedford Street and moved the buried bodies out to Queens.)[51]

In most life histories, these scenes end the story: a eulogy by Lee, a minister's assurance that "one native Indian, at least, of Oregon, is saved, as the fruit of missionary labor," and a corpse buried in a New

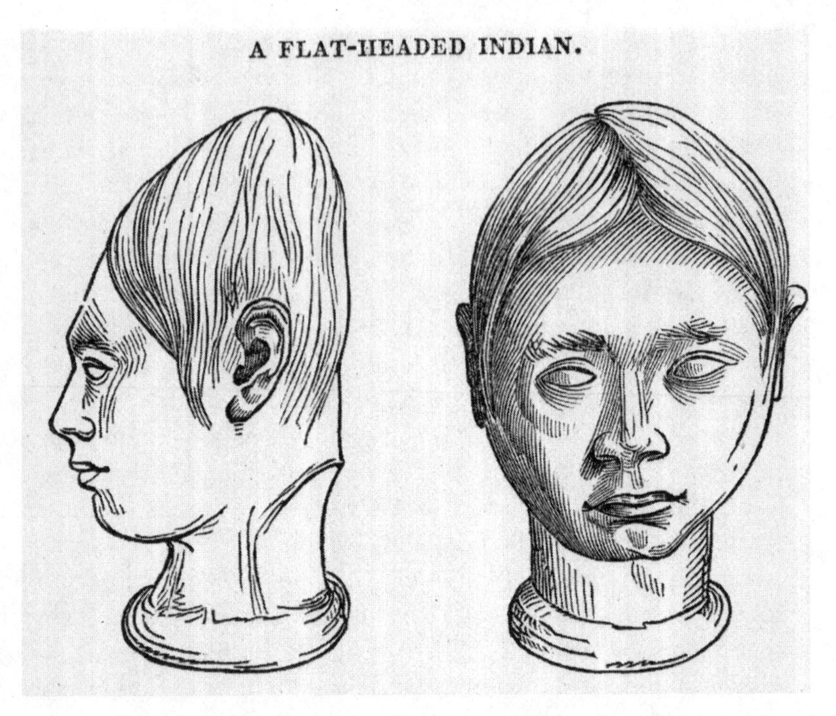

Fig. 9. "A Flat-Headed Indian," from George Combe, *Notes on the United States of North America during a Phrenological Visit in 1838–9–40* (1841).

York churchyard.[52] But given William Brooks's encounters with skull-collector Morton, his agent Townsend, and the traveling phrenologist George Combe, one has to ask at least about the material afterlife of his body. What happened to Brooks's unusual flattened skull?

Combe wanted a skull like Brooks's and, if possible, with a fresh brain. The phrenologist complained that enemies of his science kept him from getting his hands on Brooks's body, and his wish for that body gives Combe a ghoul's role in the brief biography of the man who became William Brooks. But Combe was not a villainous man. That winter Combe had come to Philadelphia to deliver his lectures and to meet Samuel George Morton, a man he knew by reputation as the country's most enthusiastic craniologist. It is not surprising that Combe encountered Jason Lee and the young men with flattened heads, who were traveling the Northeast on a lecture circuit of their own during these same months. Combe interviewed Thomas Adams in New York in

May 1839. Lee apologized that Brooks was too sick that spring to meet with the phrenologist, but the minister introduced Combe to Adams, a man from the "Cloughewallah tribe, located at the falls of the Wahlamette River," Combe wrote. (My guess is Combe tried to remember "Kalapooyah.")

Here was Combe's chance to explore what manipulation had done to his mental faculties. Combe described Adams as "intelligent, ready and fluent" on things that demanded only "observation." On questions that required "the aid of Comparison and Causality, he was dull, unintelligent, and destitute equally of ideas and language." "The organs of Destructiveness, Acquisitiveness, Secretiveness, Self-Esteem, Love of Approbation, and Firmness, were very large; those of Combativeness, Philoprogenitiveness, and Adhesiveness, deficient. It was difficult to estimate the size of the moral organs, they were so displaced," he wrote.[53]

What did Combe expect readers to conclude from this description? Adams was a destructive and secretive young man, with a good sense of self-esteem, but little love for fighting or children? Combe reassured Lee that his converts appreciated the word of God. Flattened skulls displaced but did not destroy the convolutions of the brain. The young men's brains were bent, he said, like the spine of a hunchback, and even though they had diminished capacities in faculties like comparison, they had not lost overall intelligence.

Combe's faith in his brain map seems to have blinded him to different evidence about the young men and their mental capacities, particularly their skills as observers. Accounts brim with their astute (and amusing) comparisons. Had the Scotsman listened to Brooks's words to the Methodist woman with her cinched waist, for example, he could have recorded Brooks's gift for comparative observation. He had given the woman a perfect cross-cultural comparison—waist, feet, and head. Reporters liked to record native observations of this sort. A newspaperman accompanied Lee and the "Indian youths" on a tour of New York's City Hall. "The Indians took great interest in the paintings, and made very judicious comparisons, always, however, concluding that those in military costumes were *the best.*"[54]

Combe confessed that he could not figure out precisely how head flattening changed the brain, and so he asked Lee "to carry a cast of a normal European brain with him, when he returned to his station, and to beg the medical officer of the Fur Company . . . to examine carefully the brains of these Flat-headed Indians after death, and report minutely the differences in the size and distribution of the convolutions."[55] But that spring, when he saw that William Brooks had died, the phrenologist complained that he was not given a chance to study his brain.

If Brooks had hoped that that his body remain intact, like the mummified young woman Townsend had seen in Oregon or like nearly all those he counted as his fellow Christians, then he had a little bit of luck at the end of his short, death-filled life. But it was not Lee or the Methodists who defended his remains. A doctor by the name of David Meredith Reese attended Brooks in his last illness. Physician-in-chief of Bellevue Hospital, Reese was a cranky, outspoken, obnoxious man; a passionate advocate for the American Colonization Society's plans to send freed slaves and free African Americans to settle in Africa; and a dyspeptic opponent of things he considered popular fads. He campaigned against the extreme passions he dubbed "the humbugs of New-York," including "quackery in general," "ultra temperance," "ultra abolition," and, in his mind, as bad as the rest, phrenology—a pure humbug that had somehow seduced otherwise intelligent men and women.

Reese wondered how sensible people could believe the absurd proposition that the soft stuff of the brain molded the hard matter of the skull, creating the palpable bumps that phrenologists "read" for signs of the true nature of the inner man. Adherents insisted that when "an organ increases, the skull yields by absorption at the spot against which it lies, and then, by a general growth over it, accommodates the development and displays it externally." Ridiculous, Reese said. Foolish vanity explained phrenology's appeal. It depressed him that "men and women of reason and religion, who eschew fortune-telling, witchcraft, and astrology . . . submit their own heads, and those of their sons and daughters, to these fortune-tellers, who itinerate through the country like other strolling mountebanks, for the purpose of living without labour, by practising upon public gullibility." According to Reese, phrenologists flattered clients, showering them with cheap compliments about fine character traits.[56]

The argument about phrenology protected the corpse of William Brooks. To Combe's great distress, Reese "allowed this young man to be buried without examining his brain, or at least without reporting on it, or calling in the aid of phrenologists to do so." Reese was fighting charlatans, he said. But Combe pushed back, insisting that anyone really interested in uncovering the truth would study this rare head. "It is strange that those who are so confident that phrenology is a 'humbug' should be so averse to producing evidence by which alone it can be proved to be so. The condition of the brain in a Flat-headed Indian is an interesting and unknown fact in physiology, and any medical man who has the means of throwing light on it, and neglects to use them, is not a friend to his own profession or to general science," Combe complained.[57]

If Reese was not a friend to medicine and science, as Combe thought, he did befriend William Brooks, who had come from a people with a high regard for the bodies of dead kin. Brooks's body was not handled as it would have been at home in Oregon, but it was at least interred according to the customs of his adopted religion. That must have been preferable to Brooks to falling into the hands of even so amiable a body snatcher as George Combe or so avid a skull collector as Samuel Morton.

Seeing the Dead

Why go back to William Brooks's story now? Several threads connect the characters he met on his brief transit from Chinook cradle to Christian grave. Brooks met trappers, traders, and missionaries, a herpetologist, an ornithologist, a phrenologist, and a craniologist. He spoke to Christians and politicians and posed for a portrait, which was published. Life must have surprised him. How could he have imagined a buffalo hunt on the plains, a sermon in a snowy Vermont village, a visit to a collection of "Chinese Things" in Philadelphia, rude children, curious naturalists, and staring Methodists, or his body in a grave in a small burying ground in New York's Greenwich Village? No life unfolds as a simple linear plot, but the twists in Brooks's life ask us to imagine his story as a knot formed of threads pulled from histories of biology, religion, science, and popular culture. Historians of science, technology, and society argue that we paint false pictures when we examine the individual threads separately. Power works inside the knots, at points like those in Brooks's story, where a Methodist missionary chatted with a craniologist, where serious science met popular culture.[58]

During his winter travels, Brooks met the naturalists who developed ideas about head shape, skull size, and racial difference. Excitement about his skull must have puzzled him. Why was Combe so curious about the shape of his brain? Why was Townsend so anxious to rob graves? Why was Morton certain that men with big skulls had big brains and that the race of men with the biggest brains was the best?

These men needed human skulls to spice up their stories or to prove their theories. Skull collecting also suited the morbid tastes of Morton's generation. Contemporaries wrote melancholy poems, stitched sad willows onto samplers, and spent spring afternoons wandering the paths of their garden cemeteries. They choreographed a new dance of death.[59]

Back in Oregon, William Brooks's relatives lost the ability to set death's rhythms to their cultural patterns. Epidemics killed so many so quickly that survivors could not bury the dead. Disease had ravaged a culture, fractured the community of mourners. Missionaries arrived with axes, plows, shovels, and hammers—tools to build a settlement and to convert Oregon natives into farmers. When the pace of death did not slack in the new settlements, missionaries taught pupils to make simple wood coffins so they could bury their friends

But Brooks's conversations in the East suggest that conversion to Christianity also provided emotional tools to deal with the dying, a glimpse of a future that made conversion a good choice for the orphaned youth. Brooks described faith that promised a defense against "second death." It is hard to know for sure what the phrase meant to Brooks. Did he and Lee share an understanding of salvation? Or did Brooks look to the new religion to escape devastating memories of loss?

Methodists buried Brooks as a Christian. But race trumped religion in the end. And history has brought Brooks back to us as "Stum-A-Nu, a Flat Head Boy," valued more in the end for his Indian body than for his Christian soul.[60]

1. NEGRO, born in the United States, ætat. 30. I.C. 83.

730. SEMINOLE, Warrior, ætat. 40, killed at the battle of Okee-Chobee, in Florida, December 25, 1837. I.C. 79. Nos. 726–730, inclusive, from Dr. E.H. Abadie, U.S. Army.

1524. EGYPTIAN, Memphite Necropolis. Woman, ætat. 60. I.C. 87. F.A. 79°. This is the head of the mummy opened by Mr. Gliddon in Philadelphia, January 1851, and by him presented to me.

—J. AITKEN MEIGS, *Catalogue of Human Crania, in the Collection of the Academy of Natural Sciences of Philadelphia* (1857)

CHAPTER 3

Crania Americana

"With Compliments of the Author"

Morton's work surfaced in Charleston, South Carolina, in the spring of 1839. Charleston is beautiful today—a small city that preservationists seem to have caught in a time warp—but its past was less lovely. The city sits on a spit of land between the Ashley and Cooper rivers, overlooking an Atlantic harbor that once teemed with traffic in slaves and in the rice and indigo that slaves produced. Profits from slaves' labor paved Charleston's streets and paid the architects and builders who designed and built beautiful buildings, where white Charlestonians ate and drank, danced, listened to music, and, like colleagues in Philadelphia, speculated about nature. Charlestonians made news with the country's first museum of natural history, the first American performance of Christoph Willibald Gluck's *Orfeo ed Euridice*, and with some rare early experiments in animal behavior. Thanks to Charleston's naturalists, we know that black vultures locate carrion by sight not smell. Experimental research surprised many American naturalists, who were still working (like Morton and his friends in Philadelphia) to collect, describe, and catalog New World nature.[1]

In the early decades of the nineteenth century, like their naturalist colleagues in Philadelphia, Charlestonians began to speculate about the natural history of man and about that natural history's expression in what they observed as racial differences. The city was their lab. More than half of Charleston's residents had been born in Africa or had a great-grandparent, grandparent, or parent who had been trafficked from Africa to America. The city's black majority, slave and free, frightened white Charlestonians. When tensions rose, they put streets under a curfew and sent black men and women indoors by nine each night. Charleston's mix of money, curiosity, and instability made it just the place for Morton to peddle a book on the natural history of racial difference.

We know that he tried. In the spring of 1839, Charleston's *Southern Patriot* began running ads for Dr. Samuel George Morton's *Crania Americana; or, A comparative view of the skulls of various aboriginal nations of North and South America, to which is prefixed an essay on the varieties of Human Species*. Bookseller Mr. W. H. Berrett didn't yet have finished books, but anxious customers could come by his shop and look at sample pages of Morton's great atlas of American skulls. The book would be large, a folio, printed with "new and beautiful type," on fine paper. It would include a map of the world, with continents color-coded by race, and sixty or seventy lithographs of skulls, near life-size and done with "great fidelity and elegance" by professional artists. Mr. Berrett would be happy to reserve copies of books to be shipped that summer.[2]

Morton promised to describe skulls that had belonged to people who lived in North and South America. He borrowed observations and comments from travelers, explorers, missionaries, and skull collectors. He began with heads of ancient Peruvians and then worked his way north through Brazil and Mexico and into North America. He described each of the forty groups whose skulls he had on hand, a collection that mixed remains dug from ancient burial mounds and recent dead from battles of the Seminole wars. He illustrated sixty-nine skulls (including the Seminole skull punctured by a U.S. soldier's bullet) and added a picture of an embalmed head from a cemetery in Aríca, Peru. Matthias Weaver's lithograph of artist John Neagle's portrait of Ongapatonga (Big Elk), chief of the Omawhaws, opened the book—one living man in a book of the dead. (Morton liked this portrait, describing it as embracing "more characteristic traits . . . as seen in the retreating forehead, the low brow, the seemingly unobservant eye, the large, aquiline nose, the high cheek bones, full mouth and chin, and angular face" than any Indian portrait he had seen.[3]

Crania Americana, the book that made Morton's reputation, was big and, at $20, pricey. (Imagine paying $500 for a book today.) Little wonder the price stopped scholars, men "not generally pecuniary either in their character or means," as one complained to Morton. Morton countered that the book should appeal to "every one whose attention has been directed to the study of physical man" and to everyone interested in owning a sample of American bookmaking at its finest. In the words of a supporter: "We can conscientiously and strongly recommend it to every one who takes the slightest interest in anthropological or ethnographical investigations, and who is there that does not?"

The line brims with confidence, but the number of those anxious to read up on ethnographical investigations fell short of expectations. The intellectual climate for the book may have seemed auspicious, but signs in the economy did not favor its sales. Morton began circulating a prospectus for *Crania Americana* in the fall of 1837, just as the country slid into the worst economic depression of the early nineteenth century. He pushed it hard through the collapsing economy of 1838 and 1839, and ran up against the wake of a Seminole War so unpopular that a New York bookseller warned a would-be chronicler that his record of the war would be "still born from the press." We can find signs that even wealthy Charlestonians cut back, leaving a hint of hard times in the *Southern Patriot*'s advertising columns. Just below Morton and Berrett's announcement, Mr. A. Whitney, a Guyana trader whose business had gone slack, wanted to sell "a healthy negro MAN, about 45 years of age . . . of warranted character." "He is a very intelligent and wishes to be sold to some person in the city on a wharf or as a labourer."[4]

Mr. Whitney's man probably found a new master, but Morton found very few book buyers. Historians turn easily to *Crania Americana* as a dependable cornerstone in the intellectual history of scientific racism. However, Morton's struggles in the book market ask us to take a closer look at the paths his ideas followed out of his Philadelphia study. The book, while not exactly stillborn, was not an effective vehicle for Morton's theories. Morton came to his ideas while sitting at home in Philadelphia, pondering his collection of skulls. But for his conclusions to matter, they had to circulate, to move beyond his intimate circle of naturalist friends. We think of good ideas as finding their way into a natural stream of influence. The history of *Crania Americana* is more complicated. Tracing its circulation teaches us something about how ideas make their way around an intellectual network.

The economy worked against *Crania Americana*, but Morton and his book had some things working for them. Although Morton did not present his arguments about racial difference as a defense of slavery, sharply intensifying debates over slavery helped sell the book, which offered material useful for slavery's defenders whether in Morton's Philadelphia or in the slaveholding South. The intellectual climate seemed favorable during the 1830s and 1840s when William Lloyd Garrison and his abolitionist coworkers stepped up the moral pressure on slaveholders and their allies. Morton sketched a racial order prescribed by nature, he said, and as unchanging as a shelf of dried skulls. Impecunious scholars disappointed him, but slavery's defenders who had cash to spare subscribed to the book and to Morton's ideas about racial hierarchy.

Crania Americana's introductory essay, "The Varieties of the Human Species," hints at some of the book's implications and advanced the idea that each race had been created expressly for a given geographic region. This book focuses on the two continents of the Americas, a region inhabited by people "marked by a brown complexion, long, black, lank hair, and deficient beard. The eyes are black and deep set, the brow low, the cheek-bones high, the nose large and aquiline, the mouth large, and the lips tumid and compressed." Their skulls were small; they were "averse to cultivation, and slow in acquiring knowledge; restless, revengeful, and fond of war, and wholly destitute of maritime adventure."[5] Morton imagined that his phrases described people native to Canada and to Patagonia, ancient Peruvians and contemporary Seminoles. The Eskimo, he admitted, did not fit in his American race. Eskimos must be Asians, he thought.

We credit Morton with the notion that big-headed Caucasians (remember the cannibal Pierce and the dissipated Dutchman) had big brains, but the lessons he presented in *Crania Americana* turned on arguments about history as well as skull size. He divided the American race into two families: "Demi-civilized" people of the "Toltecan Family," who had built the Mounds of North America and the architectural masterpieces of Central and South America, and the barbarous nomads of the "American Family," who chased the Toltecans from their homes and fathered the contemporary tribes of North America. *Crania Americana* weighed in on long-running discussions about the ancient history of the Americas. It is unlikely that the book's slaveholding readers spent sleepless nights puzzling over questions about the ancient mounds of the Mississippi Valley, but Morton's authoritative note on ancient history rang through his speculations on contemporary race relations. He concluded that the American race was "indigenous to the American continent, having been planted here by the hand of Omnipotence."[6]

Morton hesitated to call himself a "polygenist," a word coined in the first half of the nineteenth century to describe those who believed human races resulted from separate acts of creation, but he edged toward that position when he began to argue that each of the human races represented a different species of humanity. Like "the great majority of naturalists" of his generation, he "believed that species were immutable productions and had been created separately," to borrow Charles Darwin's language from the second sentence of his preface to *On the Origin of Species*. But Morton's conventional views about differences among species took on ugly connotations when he applied them to men and circulated them into a society that enslaved Africans. As Morton wrote: "I may here observe, that whenever I have ventured an opinion on this question, it has been in favor of the doctrine of *primeval diversities* among men,—an original adaptation of the several races to those varied circumstances of climate and locality, which, while congenial to the one are destructive to the other; and subsequent investigations have confirmed me in these views." His reasoning put a divine stamp on difference and inequality.[7]

Polygenists struggled to square their readings of history and biology with their reading of the book of Genesis, and friends warned Morton that his position could offend clergymen. Whether they believed man had descended from one pair or many, craniologists could argue for Caucasian superiority. A monogenist like Blumenbach believed his skulls proved that men of all races had descended from Adam and Eve, although descent had pulled some races lower than others. Polygenists held the harsher position that inequality was fixed by the Creator, and that neither climate nor diet could account for differences among men. Morton posited a Creator fashioning five mating pairs of humans—parents for the American, the Mongolian, the Malay, the Caucasian, and the Ethiopian races. Toward the end of his life, Morton speculated that each of these separately made races represented a different species of humanity.[8]

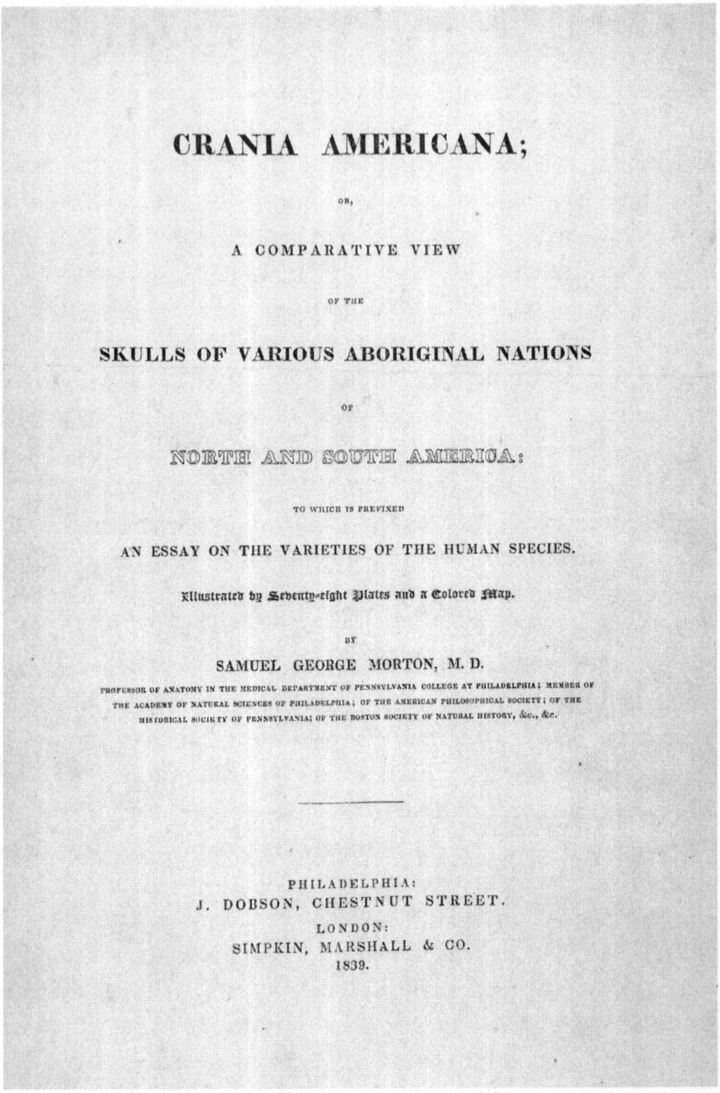

CRANIA AMERICANA;

OR,

A COMPARATIVE VIEW

OF THE

SKULLS OF VARIOUS ABORIGINAL NATIONS

OF

NORTH AND SOUTH AMERICA:

TO WHICH IS PREFIXED

AN ESSAY ON THE VARIETIES OF THE HUMAN SPECIES.

Illustrated by Seventy-eight Plates and a Colored Map.

BY

SAMUEL GEORGE MORTON, M. D.

PROFESSOR OF ANATOMY IN THE MEDICAL DEPARTMENT OF PENNSYLVANIA COLLEGE AT PHILADELPHIA; MEMBER OF THE ACADEMY OF NATURAL SCIENCES OF PHILADELPHIA; OF THE AMERICAN PHILOSOPHICAL SOCIETY; OF THE HISTORICAL SOCIETY OF PENNSYLVANIA; OF THE BOSTON SOCIETY OF NATURAL HISTORY, &c., &c.

PHILADELPHIA:
J. DOBSON, CHESTNUT STREET.
LONDON:
SIMPKIN, MARSHALL & CO.
1839.

Fig. 10. Title page to Samuel George Morton, *Crania Americana* (1839).

Morton congratulated himself on taking a contrary position. Polygenism set him in line with history's heroic dissenters. With Galileo, he thought. As Morton scribbled in the margins of his copy of *Crania Ægyptiaca*: small-minded men dismissed Galileo and, more recently, geologists "lest they cease to sustain a dogma." "And Ethnography, last of all, comes in for its share of obloquy and malediction, because all the diversified and multitudinous races of mankind, cannot be referred [he crossed out "traced"] to [a] single pair which had its cradle in Mesopotamia!" With a genealogy running back to Galileo, Morton carved himself a hero's place in the history of serious science. Friends confirmed his opinion, urging him to press on despite the "theological storm your views will bring about your ears."[9]

The book's ideas may have been scattered, but Morton found a perfect title to hold them together. *Crania Americana* pays homage to Linnaeus's binomial nomenclature, promising a descriptive taxonomy of skulls, a significant contribution to the long Enlightenment task of cataloging the natural world. Morton considered "Liber Craniorum" and "Decas Craniorium," but dismissed them both when he found a title that better captured his subject and expressed the cultural nationalism behind it: *Crania Americana* was a book about the Americas that had been made in America. And the words *Crania* and *Americana* make an especially euphonious pair, an anagram of sorts, with the letters *c-r-a-n-i-a* reappearing in *Americana*.[10]

Illustrations also added to the book's appeal. *Crania Americana*'s lithographs of delicately shaded, life-size skulls are striking still. Circumstances favored Morton, as he commissioned drawings from Philadelphia's deep pool of artistic talent. The city's reputation as the center of American lithography attracted artists from around the country. In the 1830s, authors, publishers, and printers paid artists for illustrations, for cityscapes, for decorative flourishes on government treaties, and for scientific illustrations for texts on botany and anatomy. As Morton planned his book, he had a choice of work-starved artists, happy to accept relatively steady skull work. "Beautiful day drawing skulls, mostly done," an artist reported on a good day. But sketching skulls could also be a "dry business," and Morton a fussy customer. He sometimes rejected drawings with "crabbed humor," although on another occasion he forgave this same artist when he dropped and broke a skull. Good luck brought Morton a talented man named John Collins, who drew the skulls for *Crania Americana*'s plates. Collins prospered for a time working for Morton but left contemporaries with memories of a deeply unhappy man. "Poor man, I pity him," an acquaintance wrote, "a man of education and some talent but wholly incapable to buffet the wars of adversity." Drawing skulls darkened Collins's already somber mood, and about the time he finished work on *Crania Americana*,

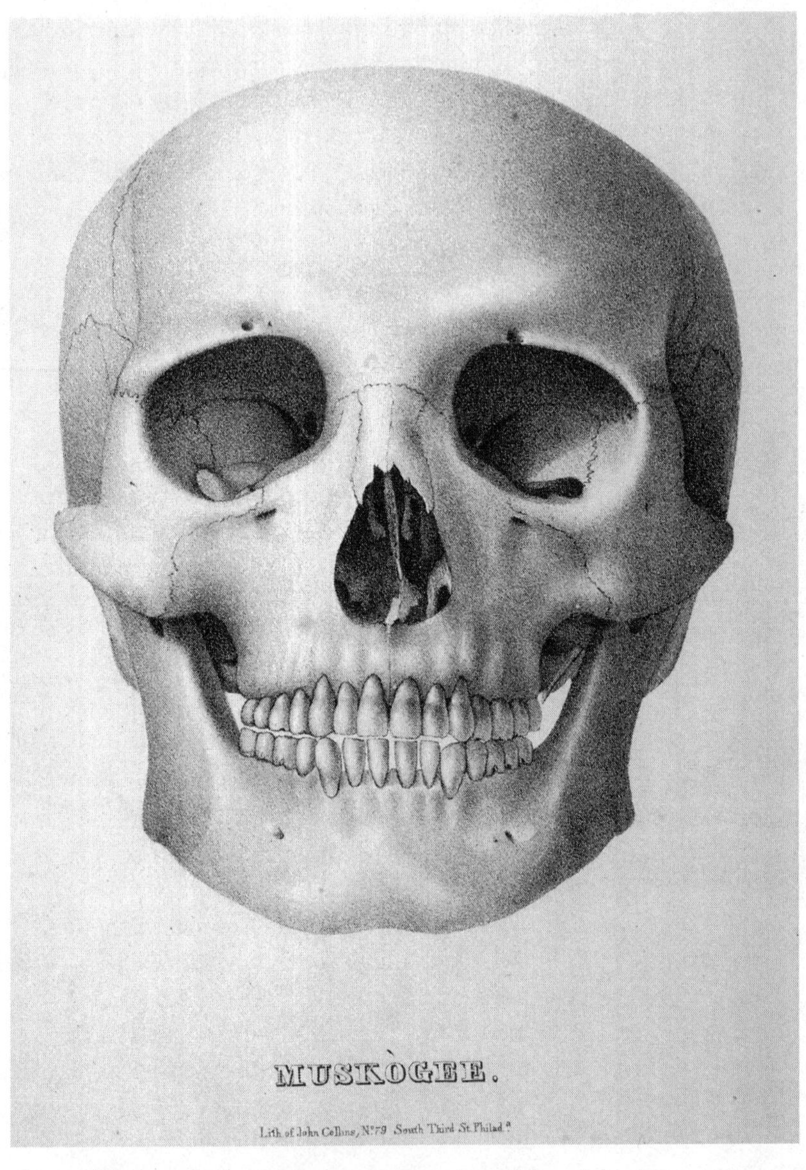

Fig. 11. "Muskogee," John Collins, lithographer, plate 26, from Samuel George Morton, *Crania Americana* (1839).

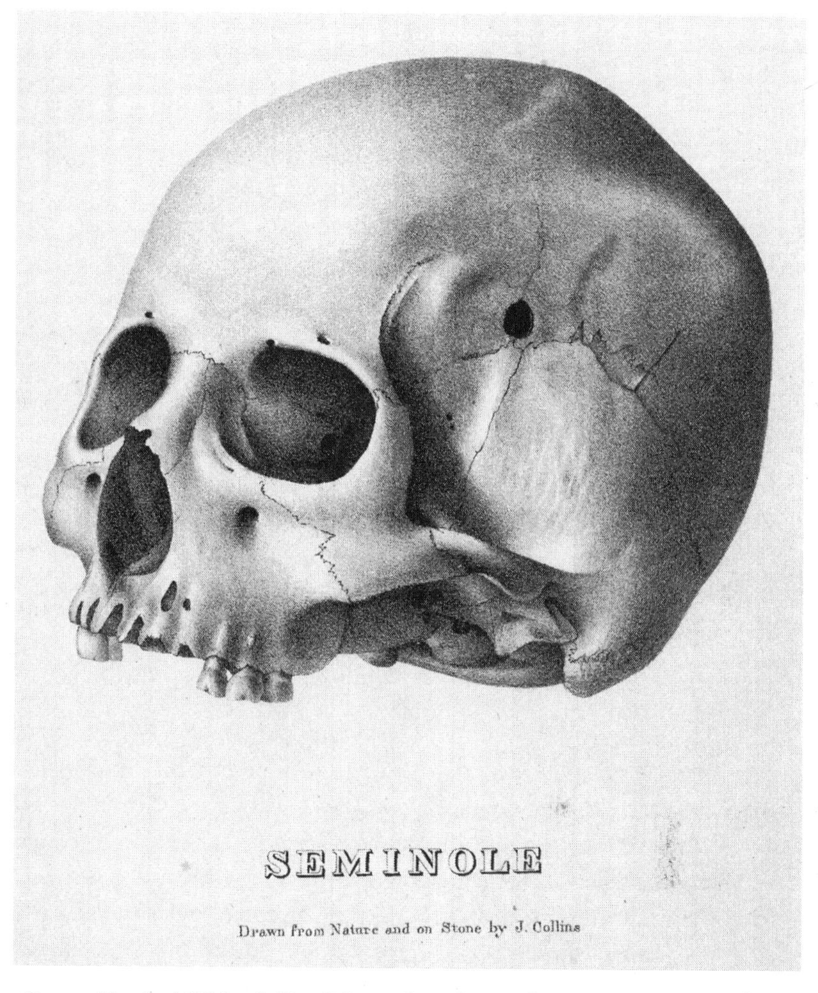

Fig. 12. "Seminole," John Collins, lithographer, plate 22, from Samuel George Morton, *Crania Americana* (1839).

he sold his lithography press and disappeared from the Philadelphia art scene.[11]

The dark humor, which colors *Crania Americana*'s lithographic plates, lent them subjective power but created problems for Morton, the scientific naturalist. As though to counter the mortal associations conjured by Collins's pictures of skulls (to make sure readers saw specimens of races and not death's-heads), Morton inserted two hundred "minor

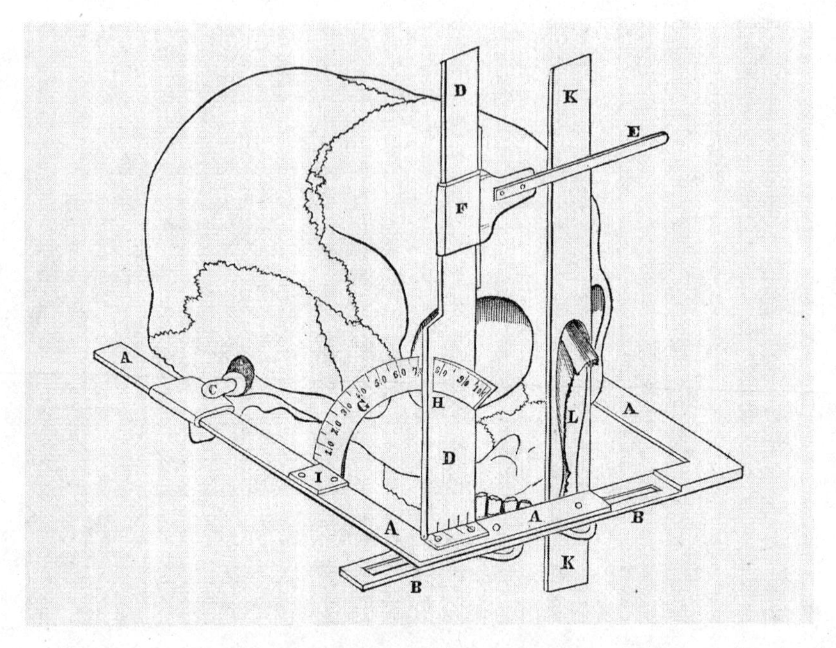

Fig. 13. "Facial Goniometer," from Samuel George Morton, *Crania Americana* (1839).

illustrations," small woodcuts of the instruments he used to measure and sketch skulls and simple line drawings that capture the prominent features of individual skulls. Morton's skull images addressed an audience schooled to see intimations of mortality in skulls. But the book countered those intimations with crania drawn "with great celerity and correctness," "by means of an instrument adapted to that purpose." To make "objective" arguments about the differences among races, the book needed to sever readers' associations between the skulls depicted in *Crania Americana* and the human relics they might have seen in paintings (or in reproductions of paintings) by Caravaggio, Hans Holbein, Frans Hals, or the Dutch masters of the *vanitas* still life. Yet even as Morton disavowed his connections to Western art's long history of representing human skulls, he drew on that history's cultural power to lend his book a touch of death's certain authority.[12]

But Morton could not take cultural authority to a Philadelphia bank, and through the early months of 1840, the costs associated with the book troubled him. The book's production costs had eaten up money he earned from fee-paying patients, but without wealthy patrons or government support, he needed to find subscribers to recover his costs. It didn't seem to be working out as he had planned. Discouraged, he added up the "years of toil and anxiety" he had spent on his skull book and wondered if a man of his "limited means" should have undertaken such a "costly work." Readers curious about "anthropological or ethnographical investigations" simply had not materialized. By 1840 only fifteen subscribers had signed up to buy *Crania Americana*, far too few to underwrite the five hundred copies he planned to publish. (Advance subscription was a customary practice for publishers and self-publishing authors looking to raise cash to produce their books.) Even the local market disappointed Morton. A sympathetic friend recalled that *Crania Americana* simply "did not, in his native city, produce the impression, and meet the success which, from a liberal spirit among the community, might have been expected."[13]

Bad luck dogged the book through the sluggish market of the 1840s. Things hit a low point in January 1840 when forty copies shipped to London disappeared when the brig *Palmer* sank in Boston Harbor. How could Morton reimburse collectors who shipped him skulls, or pay artists, lithographers, copyists, printers, and binders who contributed to the book? Friends rallied him with stories of struggling scientists. Remember Ben Franklin, the Dutchman Doornik had written some years earlier, flattering Morton that he shouldered Philadelphia's venerable intellectual traditions. A good self-image hardly helped when artists and collaborators stopped Morton on the street, asking to be paid.

Morton asked his friend the phrenologist George Combe if he should sell the skulls to pay his debts. Not a good idea, Combe replied. Not in a depressed skull market. "They are well worth £500, but I do not know who would purchase them in England or Scotland. All the great men of science who manage our public collections of natural history are hostile to phrenology, and would rather pay money to keep your collection away from the public eyes, than to acquire it, and the Phrenological societies are destitute of funds."[14]

Then quickly, in the summer of 1840, Morton's fortunes changed. James Morton, the rich uncle who had covered his school bills in Scotland, came through again. Nephew Morton kept up connections with this uncle, dedicating half the copies of *Crania Americana* to "My Dear Uncle. The Paternal kindness with which you have ever regarded my course through life, and the solicitude you have expressed for the successful completion of this work, prompt me, with feelings of mingled pleasure and gratitude, to inscribe it to you."[15] The inscription to James Morton proved wise. In the midst of his woes, Morton heard that this generous uncle had died and left him a legacy large enough to cover his debts. "This is like a scene in a play or the winding up of a novel," Combe wrote. "It is so appropriate and pleasing. How very rarely do men of talent devoted to science enjoy the good things of this life."[16]

The inheritance eased the pressure on Morton, but it did not improve the sales of *Crania Americana.* Since there was no market, Morton used copies to secure a place in the "cosmopolite fraternity" of science. In a neat hand, he wrote "with compliments of the author" on slips of paper, which he pasted inside copies of the book. He shipped complimentary copies to naturalists and scientists whose work he admired: to botanist-diplomat Joel Poinsett, to geologist Benjamin Silliman at Yale, and to anatomist John Collins Warren at the Harvard Medical School, who sent back his "best thanks for so valuable a present." Complimentary copies went to learned societies in Paris, London, Moscow, Berlin, Frankfurt, Copenhagen, to the American Academy of Arts and Sciences, and to the Asiatic Society of Bengal in Calcutta. Morton presented a signed copy to Combe and another to an artist he wanted to hire; he also sent a copy to Prussian naturalist Alexander von Humboldt, the most powerful figure in his intellectual world. Humboldt acknowledged the financial risks Morton had taken and congratulated him: "One cannot, indeed, but be surprised to see in it such evidences of artistic perfection, and that you could produce a work that is a fitting rival of whatever most beautiful has been produced either in France or England."[17]

Specimens and books moved around among these gracious gentlemen, along the circuits that connected Charleston to Philadelphia and linked inquiring Americans to curious men in London, Berlin, and Paris. Gratitude and friendship softened their hard science and drew the men into an affectionate circle. Art, numbers, friends, and the skulls themselves pulled the work beyond its crabbed concern with racial hierarchy. Morton had made a beautiful book ("sumptuous," said one reader[18]), and he and his friends pushed it out into the world. Contemporaries saw the "magnetic power by which he attracted and bound men to him, and made them glad to serve him." His "most winning gentleness of manner" drew "numerous collaborators" and bound them to him "as with cords of brotherly affection." Morton's affectionate collaborators included George Combe, the man who wanted William Brooks's

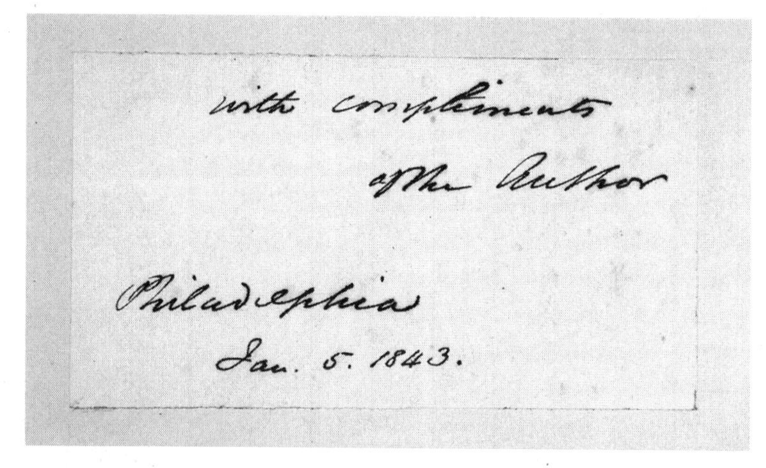

Fig. 14. "With compliments of the author," from Samuel George Morton, *Crania Americana* (1839).

brain and skull; George Gliddon, an English-born Egyptologist who unwrapped mummies for American audiences but died, drug addled, in a Panama hotel room; and finally Louis Agassiz, the Swiss-born naturalist who transformed American science from his post at Harvard. These three carried Morton's ideas out into the world and helped give them life beyond 1840s America.[19]

But there are others whose influence historians rarely consider when thinking about the Philadelphia origins of America's "scientific racism." Morton's ideas circulated into the interracial world of Charleston, South Carolina, but they originated, not just in his learned circles, but in an interracial Philadelphia, where some fifteen thousand African Americans made up around 8 percent of the city's population, the largest percentage of free blacks in a northern city. Black Philadelphians lived in neighborhoods all around the city, and Morton would have seen black faces every day. The waiter who served his lunch, the barber who shaved his face, the man who fixed his boots, the porter who delivered a box of skulls to his door, and the woman who dusted the skulls in his study could all have been African Americans. Morton shared rooms and streets and, as we shall see, a printer with African American Philadelphians. The story that surfaced in Charleston in 1839 traveled along the straight streets and back alleys of the City of Brotherly Love and out of that city in the intellectual baggage of Morton's three good friends.

George Combe (1788–1858)

The enthusiastic phrenologist George Combe (the man who wanted to study a fresh brain taken from a flattened skull) contributed an appendix of "phrenological measurements" to *Crania Americana*. His thirty-odd pages of charts and speculations followed a section on anatomical measurements. Morton and his assistant recorded their calculations of the diameter, facial angle, and cubic capacity of each skull. Combe calibrated the skulls instead for "Feelings": Amativeness (or sexual love), Philoprogenitiveness ("affection for young and tender beings"), Adhesiveness, and Destructiveness; for "Sentiments": Self Esteem, Cautiousness, Benevolence, Veneration, Firmness, Conscientiousness, Hope, Wonder ("admiration of the new, the unexpected, the grand, the wonderful, and the extraordinary"), Ideality, and Wit ("Gives the feeling of the ludicrous, and disposes to mirth"); and "Intellectual Faculties": the ability to recognize fact, form, number, order, color, to compare and to understand cause and effect. Phrenologists delighted in Combe's translation of Morton's cranial anatomy. In the long run, Combe's contribution dated Morton's work and embarrassed physical anthropologists, who have tried to understand the evolutionary significance of cranial conformations.[20]

But in the spring of 1839, Combe promised Morton something more than a scheme to interpret skulls. He could help sell the book, pitching it at lecture stops on his American speaking tour. By the 1830s, Europe's serious scientists had turned away from phrenology, but phrenology's simple schemes and promises of personal improvement still appealed to Americans, who saw advantages in being able to assess a man's character by studying his skull and liked the idea that the contours of the skull recorded a man's success in remaking himself. Morton may have remembered phrenology's iconoclastic appeal from his student days in Edinburgh and perhaps now calculated its utility to Americans dizzied by their country's demographic growth. Population doubled in the United States in the first thirty years of the century and increased almost threefold in the next thirty. Practical-minded Americans didn't have to understand the subtleties of cranial location or intricacies in debates about the material basis of ideas to appreciate tools that smoothed their way through relations with strangers—easy answers to hard questions. Is cruelty part of his character? Deception behind his plans? For painters and sculptors, phrenology's map of the skull provided a means to convey character in a few clear gestures. A high forehead. A low brow. Some students of phrenology subscribed to its tenets of self-improvement, and fingered their skulls to chart improvements as new discipline ate into the bones of old habits and appeared in new skull bumps.[21]

Morton and Combe probably shared a sense that Caucasians were the best of men, although they seem to have disagreed about connecting racial hierarchy and slavery. Morton's craniology lived on as a cornerstone of the "scientific racism" that provided intellectual support for slavery. He argued that biology, expressed in the shape and size of skulls, dictated racial destiny. Combe had more generous views on human promise, even though he hesitated to oppose slavery from his phrenological pulpits. He seems to have lacked the moral courage to risk offending potential supporters of his beloved science, though his writing preserves his outrage at America's strange contradictions. Did he have his relations with Morton in mind when he wrote: "Nothing in the United States has surprised me so much as the general tone of the public mind and the press on the subject of slavery. The institutions of America profess to be based on justice, and certainly an all-pervading justice is indispensable to their permanence and success; yet the most cruel injustice is perpetrated on the Negro race, and defended as if it were justice, by persons whose character and intelligence render them in every other respect amiable and estimable. This is a canker on the moral constitution of the country, that must produce evil continually until is it removed."[22]

Combe and Morton met during that last week of the cold December of 1838. Combe arrived in Philadelphia ready to observe the city's sights and offer Philadelphians a chance to attend his course of "phrenological lectures." He settled in at the Marshall House hotel and chased the chill from his rooms by lighting a grate heaped with Pennsylvania coal. The city's buildings impressed him, the marble-clad houses lined along the city's clean straight streets and the National Bank modeled on the Parthenon. He got a taste of Philadelphia's intellectual life when members of the American Philosophical Society and the Wistar Club invited him to dinner. He heard a good sermon by Dr. Bethune of the Reformed Church, booked a room at the Philadelphia Museum for his lecture series, and then went to visit "Dr. Morton's collection of American Indian skulls." He concluded that the collection was "valuable and extensive, and his specimens are well authenticated."[23]

Combe stayed in Philadelphia through January, pleased to sell more lecture tickets than he had in either Boston or New York. He opened the course by describing "The Brain [as] a congeries of organs manifesting different mental faculties"; expanded on "Hope, Wonder, Ideality, Wit, Imitation"; and explained the "Application of Phrenology to the present and prospective condition of the United States." The phrenological tour was going well, he thought, even though he sometimes lost audiences to local competition. To Combe's distress, managers of the Philadelphia Museum rented the floor above his lecture room to Frank Johnson and his brass band. Johnson and his African American musicians packed in white audiences. Combe counted 500 Philadelphians at his lectures but noted that Johnson's "concerts are attended by 2000 or 3000 people, the admittance being 25 cents, or one shilling sterling each. The music is so loud that it often drowns my voice, and when the audience above applaud with their feet, I have no alternative but to stop till they have done."[24]

Combe complained but the passage fits well in a book of observations on life and culture in the United States. Competition among entertainers—lecturers, musicians, dancers, actors—characterized commercial entertainment in the United States in the 1830s and 1840s. The poor Dutch Dr. Doornik with his skulls and rocks struggled to find an audience for his lectures, but he was not crazy to imagine making his living on the lecture circuit. On a given evening, Americans (segregated by race) could applaud a production of *Hamlet* (or hiss when actors stumbled drunk onto the stage). They could watch a mesmerist throw a woman into a trance, watch an automaton beat a man at chess, or attend a demonstration on electricity or daguerreotypes. They could puzzle over a "Fejee Mermaid," listen to a selection of arias, weep over the life story of a reformed drunk or a former gambler, or laugh at a performance of white minstrels in blackface. They could hear from preachers, politicians, explorers, travelers, phrenologists, abolitionists, and former slaves. Historians see these as great years for public life in the United States. But Combe knew that the good times for audiences were sometimes hard on the performers who competed for their attention.[25]

It was easier to exchange ideas in Morton's skull-filled study. Morton's collection impressed Combe. When he passed through Cambridge, Combe had seen the Native American skulls that belonged to Harvard anatomist John Collins Warren, but that collection, though "large and valuable," could not match Morton's in scale, scope, or variety. Morton explained his system for cataloging skulls: he inked a Roman numeral on the brow of each skull and then labeled each with the name of a country or tribe. He told Combe stories behind his most interesting skulls as the two swapped tips on techniques to measure skulls and exchanged recipes for things like oxygenated muriatic acid that would keep skulls from turning yellow.[26]

Combe believed he and Morton could use the collection to extend applications of phrenology from the character of individuals to the mental characteristics of nations or tribes. It pleased Combe that Morton's descriptions of the virtues and strengths of various tribes coincided with what a phrenologist would have predicted, even though, as he put it, Morton was "imperfectly acquainted with phrenology himself, and has composed his text without reference to it." "If the size of the brain and the proportions of its different parts be the index of national character, the present work which represents with great fidelity the skulls of American tribes, will be an authentic record in which the philosopher may read the native aptitudes, dispositions and mental force of those families of mankind." Combe liked Morton so much he volunteered to write some "Phrenological Remarks [that] might be suitably introduced as a Preliminary dissertation, and printed at the beginning of your Book."[27]

Better to leave phrenology in an appendix, Morton decided, although he accepted Combe's offer to help sell *Crania Americana*, and the two men grew close as collaborators. They exchanged news of wives and families as they plotted to sell the slow-moving book. (Combe's wife, Cecelia, was the daughter of the celebrated English actress Sarah Kemble Siddons. The two had married in 1833, and Cecelia brought enough money to the marriage to let Combe devote his time to phrenology.) Though made comfortable by Cecelia's money, Combe knew something about marketing books; his *Constitution of Man Considered in Relation to External Objects* ranked as one of the early nineteenth century's best sellers, running through 200,000 copies and twenty American editions between 1828 and 1860. Boston educator Horace Mann called it the "greatest book that has been written for centuries." Some feared Combe's influence matched Tom Paine's, since, like that old revolutionary, he appealed directly to readers, ignoring a need to defer to educated betters. But Combe's was a mild threat. Reformers often found his book beside those steady sellers—the Bible, *The Pilgrim's Progress*, and *Robinson Crusoe*—hardly a revolutionary's bookshelf.[28]

Combe also thought he could help Morton gain a share of the book market. He would ask Yale scientist Benjamin Silliman Sr. "to notice" Morton's *Crania Americana* in his *American Journal of Science and Arts*, one of the country's most respected scientific periodicals. Morton had already sent Silliman a copy, hoping he would review the book. It is a "splendid work," Silliman wrote in thanks, without promising anything further. Combe volunteered to write the review if Silliman would run it as an unsigned "editorial notice." Combe explained to a puzzled Morton that an "Editorial notice" would have "double the weight of a communicated one," since the notice would appear to come from the influential Silliman himself. Combe knew the power of editorial endorsements. He had used them himself, persuading newspaper editors to slip favorable comments about his lectures into editorial columns. Editorial notices sold tickets.[29]

Silliman was one of the country's leading natural scientists, not a local newsman. In 1839, when Combe and Morton approached him about *Crania Americana*, Silliman was teaching chemistry and geology to Yale students, just as he had since the turn of the century, and editing the journal he had founded in 1818. The journal often ran reviews unsigned, and Silliman accommodated Combe in what seemed a customary fashion. He ran an unsigned editorial review of *Crania Americana* in the January–March 1840 issue of the journal. The review summarized the book and praised it as "learned, lucid, and . . . classically written."

It was a bargain too, Combe wrote, "remarkably cheap, keeping in view the quantity and quality of the *materiel* of which it is composed." When Morton advertised the book, he quoted from the review, borrowing the editorial "we" to imply that Silliman had written, "We hail this work as the most extensive and valuable contribution to the natural history of man which has yet appeared on the American continent."[30] Combe's scheme left contemporary readers (and some unsuspecting historians) to assume that Silliman endorsed Morton's craniological, phrenological, and ethnological speculations. That apparent endorsement has left historians chagrined, but it would not have surprised readers accustomed to an intellectual world that did not insist on our distinctions between a Yale-based scientist like Silliman and an amateur like Morton or between popular lectures by a university chemist and those hosted by a traveling phrenologist.[31]

When even the good review did not sell books, Combe tried to rally his discouraged friend, explaining that the book "will yet recompense you, and altho' present tribulations be grievous, I sincerely trust that it is but for a season." He sympathized with Morton's losses but blamed the book's poor sales on enemies who wished to suppress phrenology's truths. When a season passed and the book still did not move, Combe suggested a new tack. "The Book is a large one & valuable and some power is needed to launch it & set it in motion. It has vitality to move onwards after that; and I am confident that it will go." He promised to "use every means in my power to recommend it, & make it known," particularly among influential friends. Combe, who had made good friends among New England intellectuals, reminded Morton that the historian George Bancroft was looking for material on American Indians. He recommended that Morton send Bancroft a copy "compliments of the author." Combe promised to take copies back to Britain and deliver them to friends, libraries, and learned societies in England and Scotland.[32]

It is easy to picture Combe and Morton bonding over human skulls and scheming to make *Crania Americana* a success. In some publications, Morton treated phrenology a bit gingerly and may have come to see his alliance with bump readers as something of a devil's bargain, but his craniology relied on phrenology's central premise that better-developed faculties show themselves by enlarging parts of the skull. In sum, intellectual power "depends upon the volume of the brain."[33] The antislavery Scotsman did not understand the United States well enough to recognize how easily Morton's arguments could be incorporated into the mid-century defense of slavery.

Like many white opponents of slavery, Combe may have believed in white superiority. But the Scotsman held a complicated sense of racial superiority and seems to have worked more easily across racial lines than his American friends. Combe might have told Morton about the "colored man" who took care of his rooms at the Marshall House. This man let Combe examine his head, as he explained how he had purchased his freedom. "Although a complete negro, [he] has a brain that would do no discredit to an European. It is of full size and the moral and intellectual regions are well developed; and his manner of thinking, speaking, and acting, indicates respectfulness, faithfulness, and reflection." It is a condescending comment, of course, and assumes that European heads set the standard. But despite his condescension, Combe developed a relationship with the family.

The man's son "Rob Roy" asked to come to Combe's lectures. How could Scotsman Combe refuse the request from a boy named for his illustrious countryman?[34] But would white Philadelphia audiences be happy to have a black man at the lecture? Combe told Rob Roy to come along but to stand near a back door, where he could hear what Combe had to say about phrenology. The rest of the audience would take him for a servant waiting for his master. Combe was willing to lecture to people of any color, provided they were "intelligent and attentive; but Americans feel differently." "I have not introduced the question of abolition into my lectures, because it is foreign to their object. So far, however, as the subject lay incidentally in my way, I have not shrunk from it, but have introduced the skulls and casts of negroes among those of other varieties of mankind, and freely expressed my opinion of the moral and intellectual capabilities indicated by their forms."[35]

Combe could have found good examples of African American intellectual capabilities right there in Philadelphia. Rob Roy and his family were part of an economically complex and culturally rich black community whose roots went back three or four generations. Since the mid-eighteenth century, slaves fleeing the upper South had settled in Philadelphia, the first free city along the Atlantic coast. West Indian families from San Domingue had been settling in the city since the turn of the century. For decades, people seeking freedom had been crossing the river from New Jersey, a state that did not outlaw slavery until 1846. Free families from the South also swelled Philadelphia's black population, especially in the 1820s and 1830s, when many moved north from South Carolina and Virginia after slave owners cracked down on suspected friends of rebels Denmark Vesey and Nat Turner.

By the 1830s, new arrivals found Philadelphia's black community well established, diverse, and cosmopolitan. Black Philadelphia had preachers, teachers, doctors, shopkeepers, and mechanics, and in numbers enough to make a middle class that supported cultural institutions like literary clubs and benevolent societies. Black property owners paid taxes and voted in elections. Unskilled black laborers dug ditches, cleaned streets, and hauled away trash, menial jobs that made the city pleasant for wealthier residents. But for many, menial jobs meant steady paying work. When the labor market collapsed with the American economy in 1837, black workers lost these jobs. In the depressed labor market of the late 1830s and early 1840s, new Irish immigrants took jobs from casual black laborers.

Black and Irish Philadelphians shared streets in poor neighborhoods like Grays Ferry, and encounters between the two groups sometimes sparked into violence. Desperately poor black Philadelphians lost jobs and homes in the shrinking economy. Reformers worried as city residents scavenged a living, peddling rags and bones picked from waste heaps. The poorest Philadelphians died on the streets or in institutions. City workers buried their bodies in mass graves in a potter's field not far from Morton's home. Morton collected local skulls from these poorest Philadelphians.

Morton's first black skulls had come from unclaimed bodies and from the institutionalized poor, reminders of poverty's racial shape. A coroner, writing in the 1840s, reported finding the poor "dead in cold and exposed rooms and garrets, board shanties five and six feet high and as many feet square, erected and rented for lodging purposes, mostly without any comforts save the bare floor with the cold penetrating between the boards, and through the holes and crevices on all sides; some in cold wet and damp cellars, with naked walls, and in many instances without floors; and others found dead lying in back yards, in alleys, and other exposed situations." A more fortunate few died in the Alms House & Hospital, where Morton and his students dissected bodies and collected their body parts.[36]

Poverty played into Morton's work. So did the social world of his racially mixed city and the growing local opposition to slavery. Morton and his naturalist friends could not have missed the sporadic racial violence that sparked through the city. In the summer of 1834, as he began working on the book that would become *Crania Americana*, white mobs attacked African Americans on the street, burned black churches, torched black-owned houses, and killed "an honest, industrious colored man" named Stephen Jones. A year later, white mobs burned houses of some of the city's wealthier African Americans.[37]

Race relations in Pennsylvania hit a low point in the late 1830s. Delegates to a Constitutional Convention stripped the vote from all black Pennsylvanians, even from longtime, tax-paying residents. But one event, the destruction of Pennsylvania Hall, came very close to Morton's door. Morton lived on Mulberry Street (or Arch Street, as it was later known) between Fourth and Fifth. Pennsylvania Hall was just around the corner on Sixth between Cherry and Race. Morton must have witnessed the spectacular work of an incendiary mob that attacked an abolitionist meeting there in May 1838 and then torched the building. Promoters built Pennsylvania Hall so "that the citizens of Philadelphia should possess a room, wherein the principles of *Liberty*, and *Equality of Civil Rights*, could be freely discussed, and the evils of slavery fearlessly portrayed," as its abolitionist sponsors put it.

Their fine vision lasted about four days. Pennsylvania Hall opened on May 14, welcoming abolitionists Angelina Grimké, Lucretia Mott, Abby Kelley, Maria Chapman, and William Lloyd Garrison, who complained to the building's managers that there was not a "colored brother" on the platform. In the audience, the races mixed. But as Combe reminded Rob Roy, "Americans" did not like this race mixing. One witness reported that it angered a white mob to see "white and coloured persons . . . leaving arm-in-arm, a sight which had never before been witnessed, in this city at least, nor, perhaps, in any other part of America."[38]

A crowd gathered during the meeting, shouting down the speakers and throwing bricks through windows. "When the meeting broke up, some of the negroes were assaulted as they came out, and the rest were enabled to escape through the back entrances." Witnesses described the mob as a mixed group, "respectable" Philadelphians joining rowdy dockworkers, rough and respectable both bent on destruction. River men came up from the docks, armed with axes and crowbars and carrying buckets of tar and turpentine to fuel fires. Reports on the work of the well-organized mob painted a sorry picture of the state of law and order in Philadelphia. Notices posted around the city called on the "friends of property" to come out and shout down the meeting. Philadelphia police knew something was brewing but did nothing.

On the night of May 17, the mob set the building on fire, kindling flames with splinters broken from window blinds. "In a very short time the whole edifice was a sheet of fire, illuminating the city far and wide with a brilliance equal to that of noon-day." Spurred by a rumor that Philadelphia's African Americans were "armed, and ready for a rising," the mob burned the Shelter for Colored Orphans and attacked a church.[39]

Even if Morton slept through the fire (difficult in this neighborhood of wooden houses), he would have woken the next morning to the acrid smell of the still-smoldering building. When Combe arrived in Philadelphia at the end of the year, the ruins of Pennsylvania Hall still stood,

Fig. 15. "Pennsylvania Hall in Flames," drawn by John Sartain, from *A History of Pennsylvania Hall, which was destroyed by a Mob, 17th of May 1838* (1838). (Courtesy of the American Antiquarian Society.)

a skeletal reminder of the country's potential for racial violence and the lost hope of a mixed-race public life. Combe could have stopped by the Anti-Slavery office on Ninth Street and picked up a souvenir print of the building. He did mention that by the spring of 1839, Philadelphians were "beginning to be ashamed" of the standing relic of their year-old "outrage."[40]

During his American travels, Combe often kept his antislavery opinions to himself. Promoting phrenology came before freeing slaves, although he did put his opinions in the published account of his "phrenological tour." He blamed Britain for saddling the American colonies with slaves, but pointed out that Americans were the ones living with the contradictions between high-minded political rhetoric of freedom and legalized bondage. As he traveled south from Philadelphia into the land of slave owners, Combe puzzled over the United States. Advertisements for slaves in Washington, D.C., papers ran alongside reprints of congressional speeches on freedom and liberty. It seemed to him that American politics was "a drama written by a madman." "Without pretending to any uncommon degree of sensibility, I confess that my mind could never look on slaves, particularly children, and young men and women, without involuntarily first placing myself in their stead, and

Fig. 16. "Pennsylvania Hall following the Fire," drawn by R. S. Gilbert, from *A History of Pennsylvania Hall, which was destroyed by a Mob, 17th of May 1838* (1838). (Courtesy of the American Antiquarian Society.)

then following them to the 'Louisiana and Mississippi market,' to the cotton and sugar plantations, where they are forced to labor to the limits of their strength, till toil and misery send them to their grave. These ideas haunted my imagination, until the whole subject became deeply distressing."[41]

Morton did not join Combe on his imagined journey. The scientific craniologist shunned Combe's romantic sympathies, just as he shunned the superstitious laymen who wondered at his taste for skulls. Morton imagined a new science cut free from the misguided sentiments of abolitionists and reformers so visible in the cultural life of early nineteenth-century America. When Morton began to voice his beliefs in polygenism and a divine ordination of racial hierarchy, he hid behind hard-headed empiricism. Craniometry simply measured things as they were, he said. His single-minded devotion to reality freed him from the misplaced sympathies of a man like Combe with his reformer's hopes for a better world. "Nothing so humbles, so crushes my spirit, as to look into a mad-house, and behold the driveling, brutal idiocy so conspicuous in such places," Morton wrote. "It conveys a terrific idea of the disparity of human intelligences. But there is the unyielding, insuperable reality." It was the task of the man of science to face reality, Morton told himself, and without a crutch of superstition or sentiment.

Combe died in 1858, on a visit to Surrey, still believing that phrenology provided an accurate map of the brain, still optimistic about human improvement, but still wondering if the United States would find its way out of slavery.[42]

George Robbins Gliddon (1809–1857)

When Combe returned to Scotland in 1840, Morton turned to new advocates with warm personalities but harsher racial views: George Robbins Gliddon, an eccentric English-born Egyptologist, and Louis Agassiz, the Harvard-based naturalist and "founding father" of modern American science. Each of these men struck contemporaries as exceptionally gregarious. They were good-natured men endowed with a special magnetism that draws and holds audiences. Salesmen, after a fashion. And their good cheer helped to smuggle ideas about racial hierarchy into American culture. Audiences who bought tickets to hear Gliddon lecture on the pyramids learned about the antiquity of racial differences; men and women who came to hear Agassiz talk about animal life on different continents picked up hints that human beings, too, might have been made as separate beings (and so not of one flesh) by a creator devising different creatures for each continent. That some human beings had come out better than others was just one of history's unhappy facts, they said.

George Gliddon was born in 1809, the eldest son of an English merchant with business dealings around the Mediterranean. George joined his father in Egypt in 1818 and stayed for the next twenty-three years. He and his father cultivated ties to the United States, accepting honorary posts as consul and vice-consul at Alexandria and Cairo. Congress did not pay a salary to these low-level diplomats but did reimburse the "official expenses" that mounted up when they took on the small problems of traders and travelers. Traffic between the United States and Egypt was sparse in the 1830s, but visitors thanked the younger Gliddon for finding porters to carry bags and guides for tours of the pyramids. He had tips for travelers knocked back by Egyptian heat. Noonday temperatures spiked at ninety-five degrees one early May, wasting visiting New Yorkers. Close the shutters, Gliddon told them, when they marveled at his cool apartments.[43]

In 1837 Mohammed Ali, Egypt's modernizing viceroy, sent Gliddon to the United States to buy machines to process the cotton, rice, and

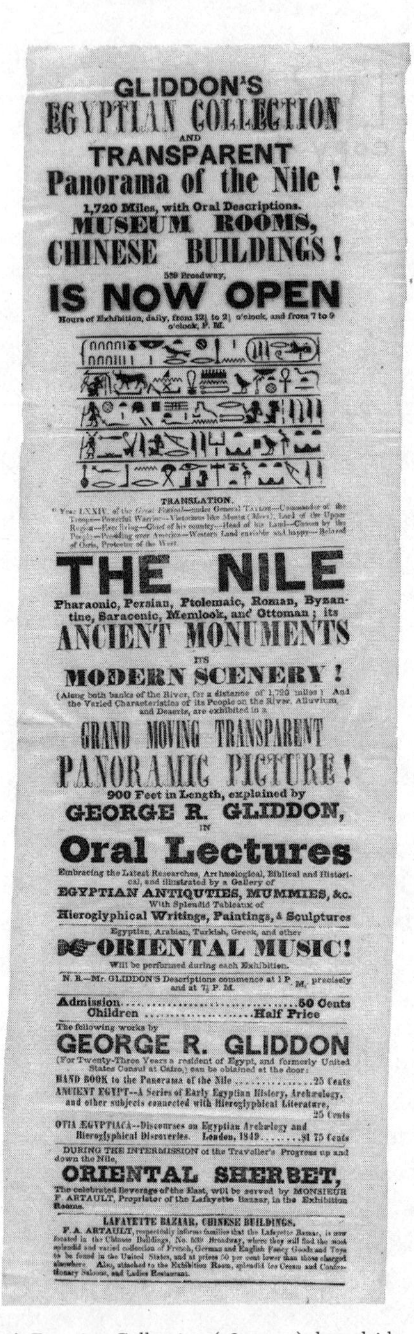

Fig. 17. Gliddon's Egyptian Collection (1850–57); broadside advertisement (Collection of The New-York Historical Society.)

sugar crops he had introduced. Over the next few years, Gliddon would sour on the viceroy's projects, complaining that the government's cotton monopoly had impoverished poor farmers and that the government's gunpowder factories destroyed historic monuments, as workers leached saltpeter from stones pilfered from ancient temples, but he seems to have taken his first American visit with earnest hope for trade.[44]

Morton heard about Gliddon on this trip. The Philadelphian needed Egyptian skulls to prove his hunch that humanity had always been divided into racial groups. In his mind, ancient Egypt was about as close as one could get to the dawn of creation. And Egypt had lots of old bodies, relics from thousands of years of embalming. There was a sprawling necropolis in the middle of Cairo, and the desert that edged the length of the fertile Nile Valley was riddled with tombs. (In an 1847 lecture, Gliddon estimated Egypt's mummy count at an astonishing 500 million, though he feared that tomb robbers had cut that number.) Morton sent Gliddon a polite note, apologizing that poor health kept him from coming to New York himself but asking if Gliddon could "employ a confidential and well-qualified person" to get him twenty-five or thirty skulls. He tucked in a prospectus for *Crania Americana* to give Gliddon an idea of his project.[45]

This work on race and skulls struck a chord, and back in Cairo Gliddon set to work on Morton's skull order. In the spring of 1839, he wrote to Morton that he had been "collecting and putting by without thinking of the number" until he had "93 skulls in my house!" "93 relics of humanity 'grinning horribly their ghastly smiles,' out of Cupboards, Boxes, and Shelves, to the horror of the only servant, who is admitted to this Sanctum, and who is greatly scandalized at his master's abomination." No doubt, poetic Morton caught Gliddon's reference to John Milton's "ghastly smiles," and the two men shared a wink over the frightened servant. Gliddon liked this skull business. "Indeed, it has afforded me a sort of rascally pleasure," he wrote, "and I would make you laugh at my numerous experiments in the resurrection line." Gliddon found ready helpers in places like Qurna in western Thebes, where tomb robbers squatted in the dust and hawked skulls, bones, and whole mummies.[46]

Gliddon bought several dozen mummies himself, but he concentrated on crania for Morton and tried to find the full range of Egyptian skulls—ancient Egyptians, Ptolemaic Egyptians, Copts, Jews, Bedouins, peasants, Arabs, Mamalouks, Greeks, and Armenians. Good news for Morton, on the one hand, but he could not have been happy to see the bills mount up. Gliddon paid one man $7.50 a month to collect skulls from the island of Philae and from Nubian cemeteries in "Upper Egypt." He hired a "snake hunter" to help exhume skulls from Cairo's sprawling necropolis. He told Morton that he and the snake hunter shared "many a chuckle" as they "abstract[ed] skulls from Convents, Tombs, Sanctuaries, and mummy pits." But that snake hunter had charged Gliddon $5 a month and demanded an extra $25 as "compensation for the risks encountered if he had been discovered by the Police." And then Gliddon had had to set aside additional money for "backsheesh" to pay off the police who did discover them. The costs added up. A year's collecting left Gliddon out $83.65. He needled the doctor to pay, trusting Morton to cover the $100 note he had drawn from a visiting New Yorker. (The extra $16 would go to completing the collection, he promised.) He worried about getting skulls through customs at Alexandria and finding a merchant captain brave enough not to "throw them overboard *in a gale of wind*." By the time he left Egypt in 1841, he had shipped Morton more than one hundred skulls, making him the single most generous of suppliers.[47]

That summer he wrote to Morton from London, inquiring about the Egyptian skulls and asking about the balance on the note he had taken out in Morton's name. He threw in a little flattery. On a table at the Royal Geographical Society, he saw "for the first time 'Morton's Crania Americana'; which electrified me, and has excited in me a sort of craving to hear from you some account of what you think of my skulls." Morton, who received so many letters, had been slow to answer Gliddon, and Gliddon would not get Morton's opinion on the skulls until he arrived in the United States in January 1842, ready to join America's bustling lecture circuit, offering a course "on Egypt and its monuments."

Morton gave Gliddon a "friendly welcome," and the two went to work on Gliddon's Egyptian skulls, using them to bolster arguments about polygenism and the antiquity of racial hierarchy. Skulls, mummified heads, reliefs in temples and tombs, all taught them that "the physical or organic characters which distinguish the several races of men are as old as the oldest records of our species." And when they studied Gliddon's sketches of Africans taken captive by victorious pharaohs, they concluded that "Negroes were numerous in Egypt, but their social position in ancient times was the same that it now is, that of servants and slaves." With that, they dropped a mantle of Egypt's antiquity over America's contemporary race making.[48]

Morton published their theories in 1844 in *Crania Ægyptiaca; or, Observations on Egyptian Ethnography, derived from anatomy, history and the monuments*, dedicating the book to Gliddon as "a memento of the esteem and friendship of the author." It had the imprint of the American Philosophical Society, the venerable organization that carried Ben Franklin's intellectual legacy into the nineteenth century. Society members heard Morton's speculations about Egypt at a meeting on May 29, 1843, when he carried twenty-nine embalmed heads to the society's hall and "submitted them to the inspection" of his colleagues.

Despite the venerable imprint, *Crania Ægyptiaca* was a modest production, running just sixty-seven pages. Artist Matthias Weaver drew its twelve lithographic illustrations, and the gifted Richard Kern sketched portraits of the kings and queens of Egypt.[49] Weaver complained that the book's tight budget forced him to cram lithographic stones with small drawings of leprous-looking mummy heads. The art would never match *Crania Americana*, but with Gliddon's push the book sold well. He took copies to Charleston, where he knew the book "would draw plenty of customers" among pro-slavery intellectuals and their political allies ready to credit Caucasians for Egypt's accomplishments. (Abolitionists argued the opposite, describing Egypt as an African civilization whose accomplishments had surpassed the white civilizations of Greece and Rome.)[50]

But it was Gliddon's successful turn on the lecture circuit that brought these ideas to Americans who had never heard of the American Philosophical Society and cared little about the opinions of Philadelphia's learned gentlemen. Gliddon's lectures on Egypt carried the book's racist arguments to audiences caught up in an American wave of Egyptomania. Bonaparte and his savants launched the craze for Egypt early in the century, when they carted Egyptian treasures back to Paris. Jean-François Champollion fueled it in 1822 by deciphering the hieroglyphs on the Rosetta Stone. Scholars stood ready to open whole new chapters on the 5,000-year-old civilization. And Gliddon brought it all to America, riding the wave of Egyptomania around the United States in the 1840s. Audiences in Boston, New York, Philadelphia, Baltimore, St. Louis, Pittsburgh, Cincinnati, Chillicothe, Portsmouth, Savannah, and Charleston turned out to see his 800-foot "Grand Moving Trans-

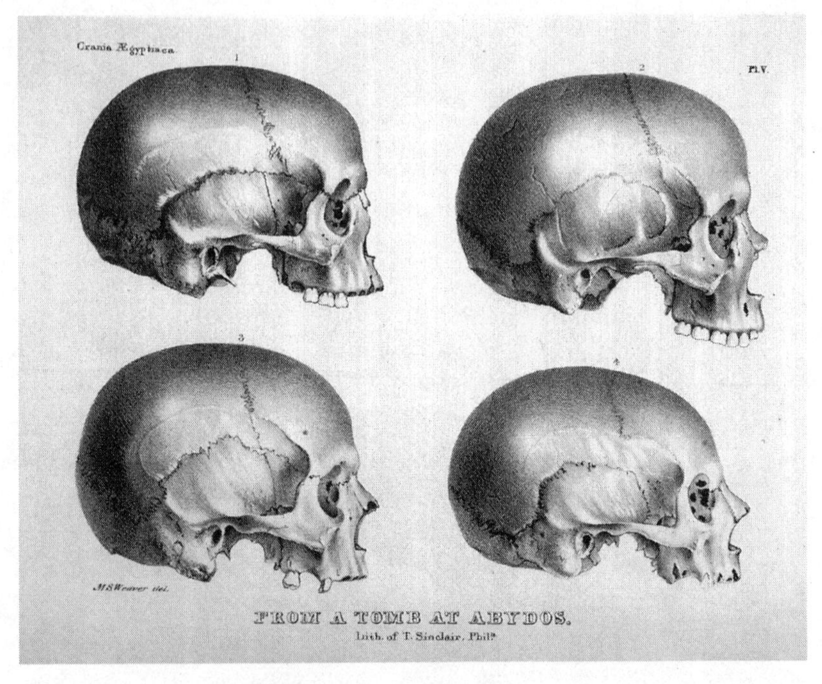

Fig. 18. "From a Tomb at Abydos," Lith. of T. Sinclair, M. S. Weaver, del., from Samuel George Morton, *Crania Ægyptiaca* (1844).

parency." He led armchair tourists up one bank of the Nile and down the other, lecturing on "Early Egyptian History, Archaeology, and Other Subjects Connected with Hieroglyphical Literature."[51]

Gliddon could sound like a professor, delivering lectures in a strange polysyllabic vocabulary, peppered with references to scholarship in Hebrew, Greek, Latin, and French. And he looked like a preacher, dressed in black, standing on a stage between two tables, one piled with scholarly books, the other with pots and scarabs he had smuggled out of Egypt. On good days he swept his audiences into a "magic region," where Bible scenes came alive. Noah's son Ham brought his family into Egypt's desert; Joseph whispered advice to the pharaoh; little Moses squirmed in the bulrushes; Mary and Joseph hid the baby Jesus from Herod's executioners. Gliddon's firsthand descriptions of the Sphinx, the Pyramids of Giza, and the Colossi of Memnon, of crocodiles, *Sic-sacs*, hippos, pelicans, vultures, ostriches, water buffaloes, and even the world's largest sycamore tree (the wood of choice for a mummy's

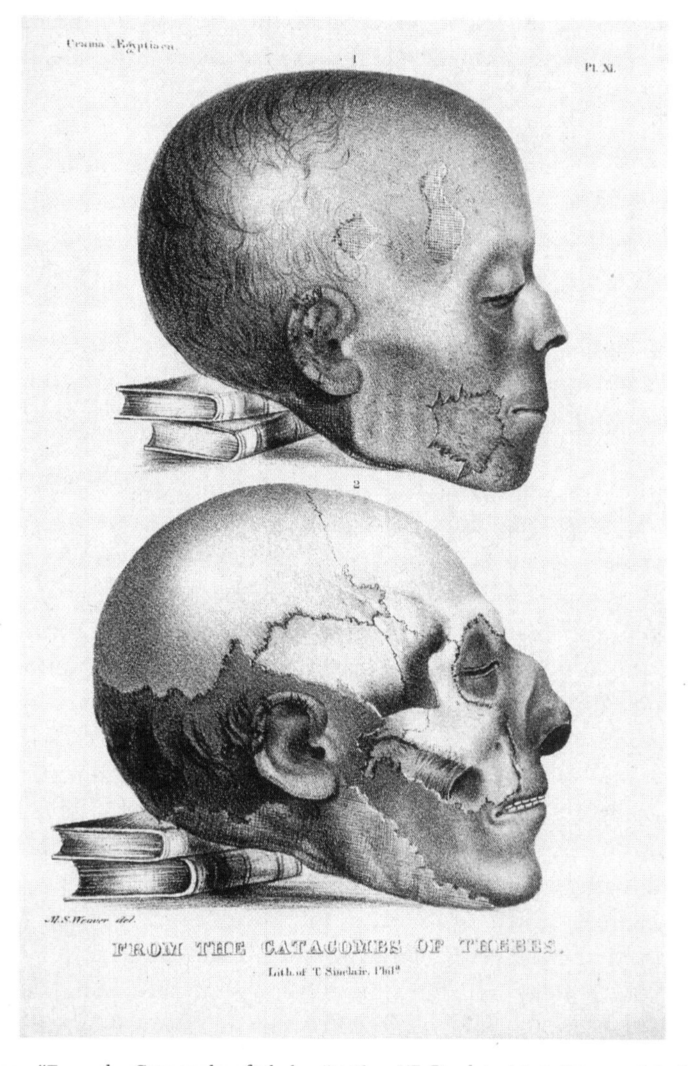

Fig. 19. "From the Catacombs of Thebes," Lith. of T. Sinclair, M. S. Weaver, del., from Samuel George Morton, *Crania Ægyptiaca* (1844).

coffin), made his history lessons especially vivid. He saved a last lecture for "reflections on the historical illustrations of the families of Man on Ancient Egyptian Monuments, by the Author of 'Crania Americana.' "[52]

Gliddon might have disappeared into the footnotes on history's eccentric entertainers, but for the fact that he actually represented features of the scientific establishment of the 1840s. For a decade he maintained his balance on the line where serious learning bumped into popular entertainment, taking cues from both Louis Agassiz and P. T. Barnum.

In 1843 Boston's intellectual establishment invited Gliddon to deliver a course of Lowell Lectures on Egyptian archaeology. These lectures were the brainchild of wealthy Boston cotton manufacturer John Lowell, who died in 1836, leaving money in his will to endow public lectures that brought together theology and science. Gliddon followed distinguished predecessors, including Yale's Benjamin Silliman, Massachusetts governor Edward Everett, and naturalist Thomas Nuttall. He speculated about the age of Egyptian civilization, translated some hieroglyphs, and tested the racial theories that he and Morton had conjured from their skulls and mummy heads.[53]

But respectability proved challenging for Gliddon. New York's lively press corps caught another side of the man in 1843 when he got into a fight with James Ewing Cooley, an auctioneer turned satirical travel writer. Gliddon trashed Cooley's *The American in Egypt with Rambles through Arabia, Petraea, and the Holy Land during the Years 1839–1840* in the *New York World*, denouncing the book as a slander on his father. Cooley described the elder Gliddon as a pompous, lazy man, slinking around a flea-infested apartment in his Turkish slippers. As an American consul, he was useless. Cooley was "anything but a gentleman," Gliddon wrote. The mild rebuke provoked Cooley, who followed Gliddon into a store on lower Broadway and punched him in the face. Breaking through the laughter in the courtroom, the jury found Cooley guilty of assault and the judge fined him $5, a slap on the wrist to acknowledge that circumstances had "partially justified" the attack.[54]

Gliddon had a harder time escaping from the laughter around a second incident, when he unwrapped a mummy in Boston in June 1850. Wealthy Londoners had developed the pastime for mummy unrolling, and Gliddon brought it to the States. Curious Bostonians, clutching their $5 tickets, crowded into Tremont Temple for the three-day mummy event. The "elite of Boston and Cambridge" turned out, including poet Henry Wadsworth Longfellow, Harvard president Jared Sparks, and his Harvard colleagues—anatomist Dr. Oliver Wendell Holmes, obstetrician Dr. Walter Channing, botanist and architect Jacob Bigelow, and naturalist Louis Agassiz, who came on stage to inspect the linen mummy cloth. Gliddon planned for suspense to mount as he peeled back each of the layers that concealed the mummy. On Monday Gliddon chattered about the Nile and the Pyramids at Giza while a carpenter sawed through a sycamore mummy case. On Wednesday he began to unwrap the mummy, the daughter of a priest, he said. (With the help of the local press, the priest's daughter morphed into a mummified princess.) But at Friday's last unwrapping, the princess proved to be either a prince or a priest's son. Gliddon blamed a coffin maker's sloppy hieroglyphs. Papers across the country picked up the story, giving readers a chance for a laugh at Boston's best.[55]

The story left Gliddon a figure of farce, not a happy thing for a man trying to maintain a balance between entertainment and education. Even before the Boston fiasco, Edgar Allan Poe saw the absurdity in the mummy passion and took a dig at Gliddon, Morton, and the traveling British commentator James Silk Buckingham in "Some Words with a Mummy," a story first published in the *American Review* in 1845. The ancient Egyptian "*Allamistakeo*" comes back to life to mock his unwrappers.[56]

The Boston episode sent Gliddon south and into the camp of intellectual racists led by Mobile, Alabama, doctor Josiah Clark Nott, a man who speculated correctly that mosquitoes carried yellow fever, but who also pushed polygenists' theories about the separate origins of human races. With Nott's help, Gliddon continued to circulate the pernicious ideas about racial difference that he and Morton had hatched. When they learned that Morton had died, Gliddon and Nott began work on a baggy racist tome, *Types of Mankind; or, Ethnological Researches, Based upon the Ancient Monuments, Painting Sculpture and Crania of Races and upon Their Natural, Geographical, Philological and Biblical History*, a book they dedicated to Morton's memory. Harvard naturalist Louis Agassiz gave them the lead essay, "A Sketch of the Natural Provinces of the Animal World and Their Relation to the Different Types of Man," which set the stage for the book's polygenist arguments. *Types of Mankind* brought Morton's theories about skulls as a record of racial hierarchy to new groups of readers, including a handful of firebrand southerners who liked his idea that " 'Negro-Races' had 'ever been *Servants* and *slaves*.' "[57]

The book sold well, but Gliddon, lacking a writer's temperament, "wearied of sedentary life," an obituary reported. He took a job as deputy agent of the Honduras Interoceanic Railway and headed for Central America. He fought with coworkers and left Honduras for Panama City, where he died in a hotel room in 1857, the victim of an opium overdose, pulmonary congestion, or "fatal isthmus fever," depending on the account you read.[58]

Louis Agassiz (1807–1873)

The Swiss-born naturalist Louis Agassiz would help to pull Morton's work through the 1850s, smuggling its racial science past the Civil War and setting it to work in post-emancipation America. Like Combe and Gliddon, Agassiz was warm and outgoing and drawn to the quiet Morton. Contemporaries describe Agassiz with happy adjectives: handsome, impulsive, kind, generous, magnetic, enthusiastic, supportive, curious, and good-humored. His wit and his lovely French-accented English appealed to American lecture audiences, who found him a zealous advocate for scientific theories, his own and Morton's.

Agassiz met Morton in the fall of 1846, just as the Swiss naturalist was beginning the American phase of his scientific career. He arrived in the United States with a reputation as one of Europe's leading naturalists. As a young man, Agassiz had developed a passion for natural science and worked his way from his father's small Swiss parish to Europe's capitals. He studied in Paris with paleontologist Georges Cuvier, who thought the young man so gifted that he handed him notes and drawings for the work he had planned on fossil fish. Prussian geologist Alexander von Humboldt became a patron, too, and helped assure that Agassiz had financial support to pursue his interests. In 1832 Agassiz returned to Switzerland, teaching natural science at the Lyceum at Neuchâtel in Switzerland and publishing his first great work on European fossil fish. He also began to study glaciers and to develop his theory that natural history recorded abrupt changes in the earth's past. Agassiz imagined a creator staging natural catastrophes that wiped out old species to make way for new.

In the early 1840s, Agassiz was run ragged by his writing and teaching and by his failing marriage. He accepted an invitation in 1846 to come to the United States and deliver the Lowell Lectures on "The Plan of Creation in the Animal Kingdom." He arrived in Boston in October 1846, with two months to prepare his lectures. He found the country a match for his energy. The United States abounded with things to study and men willing and able to help him collect and study them. Everything surprised him—the landscape, the trees, the fast trains, the traces of glacial moraines, and even the 3,000 neatly dressed working men meeting in Boston to form a library. He toured the Northeast, stopping to see naturalists in New Haven, New York, Princeton, Washington, D.C., and Philadelphia. In zeal and knowledge, American scientists "seem to compete with us, and in ardor and activity they even surpass most of our savants."[59]

Morton impressed him most. On Agassiz's first visit to Philadelphia, the two spent hours together, with Morton taking him through the collections at the Academy of Natural Sciences. He described Morton as a man "entirely to his taste, a man of science in the best sense; admirable both as regards his knowledge and his activity." And Agassiz "delighted" Morton "with his astonishing memory, quick perceptions, encyclopedial knowledge of Natural History and most pleasing manner. There is no affectation or distrust about him," he wrote.

Morton's skull collection pleased Agassiz, who told his mother he thought the "collection alone worth a journey to America." Agassiz and Morton shared a passion for collecting, confidence in careful empirical descriptions and the value of comparative analysis, and faith that scientific ideas grew from work on specimens. Observe first, they agreed, and then generalize. Perhaps they compared notes on their careers, complaining about the unhappy months each had spent as a young man apprenticed to a merchant and the years in medical school that seemed a long and indirect route to the study of natural history. Morton also offered newcomer Agassiz access to the Academy of Natural Science's working network of correspondents around the country, an "army of intermediaries," one scholar has called them. Morton and Agassiz both knew that their work depended on amateur collectors who sent specimens—whether skulls or fish—to Philadelphia or to Cambridge.[60]

Eventually, the conversation turned to creation, catastrophe, and the races of men. They agreed that polygenism seemed the best explanation for differences among men, as it did for the differences in the flora and fauna of the continents. Before coming to the States, Agassiz had described the races of men as families of a single species, but after meeting Morton, he began to imagine races as different species, each specifically created for one of the earth's continents. Experiences in a racially diverse America also seem to have encouraged Agassiz's move toward polygenism. Like many educated white contemporaries, Morton and Agassiz did not support slavery, but they did not espouse racial equality. Scientific inquiry floated above the tangled politics of slavery and abolition. "Science had nothing to do with such an iniquity; to deal with it was the work of morality, philanthropy, politics, and religion, but not of the savant whose domain is entirely outside of all institutions of society" was how an early biographer characterized Agassiz's feelings.[61]

Pleased by the meeting, Morton gave Agassiz a copy of *Crania Americana*, another volume from the unsold stock, and Agassiz carried the heavy book back to his Philadelphia hotel. At the hotel, an encounter with African American waiters disturbed him, deeply. Like Combe, Agassiz noticed that men with black skin staffed the hotel, cleaned his rooms, and waited on tables. But Agassiz did not ask about the history or the families of hotel workers, as Combe had done when he chatted with Rob Roy and his parents. Instead, he wrote to his mother, expressing shock at his first close encounters with men of a different race. The letter is painful to read.

"All the servants at the hotel I stayed in were men of color," Agassiz wrote.

> I scarcely dare tell you the painful impression I received, so contrary was the sentiment they inspired in me to our ideas of the fraternity of humankind and the unique origin of our species. . . . Seeing their black faces with their fat lips and grimacing teeth, the wool of their heads, their bent knees, their elongated hands, their large curved fingers, and above all the livid color of their palms, I could not turn my eyes from their face in order to tell them to keep their distance, and when they advanced their hideous hand toward my plate to serve me, I wished I could leave in order to eat a piece of bread apart rather than dine with such service.[62]

Did Agassiz's staring pass a limit set on rudeness in the polite hotel quarters of Philadelphia's racially structured society? Or was he simply practicing the visual skills that made him a good scientist on men he regarded more as specimens and curiosities than as fellow humans? Both Agassiz and Morton bragged about their skills as observers, their ability to spot what others missed, to see beneath a surface that would distract ordinary men. Rumor had it that Morton could discern race or nationality from a cranium or reconstruct a face to go with a skull, contemplating "it fixedly in every position, noting every prominence and depression, estimating the extent and depth of every muscular or ligamentous attachment, until he could, as it were, build up the soft parts upon their bony substratum, and see the individual as in life." He once identified a Phoenician skull, though he had never seen one. "It was what he conceived that Phoenician skull should be, and it could be no other." (Stories with similar boasts circulated about naturalists who could identify a fish from a scale or, most famously, Georges Cuvier, who could name an animal from a tooth or a bit of bone.)[63] Morton and Agassiz shared confidence in their ability to see the patterns that would lead them to discern a logic and order in nature that others missed. Little matter what waiters thought about men of science staring at their faces, for what had science to do with manners or with present iniquities or present politics.

Joseph Willson and Frederick Douglass

Morton and Agassiz, even if they didn't want to talk to black men and women, could have read a short book called *Sketches of the Higher Classes of Colored Society in Philadelphia*, which was published in the city in 1841. The book's author was Joseph Willson, the well-educated son of a wealthy white Georgia banker and a free African American woman. The banker died and left an estate that provided well for his widow and mixed-race family. Willson and his mother left Georgia and moved to Philadelphia in 1833, when he was about sixteen. Willson lived and worked among the city's prosperous black residents, the "higher classes" he profiled in his book. He acknowledged that white readers might expect his *Sketches* to be a burlesque, a minstrel's mockery of black striving, but mockery was not what he had mind.

Willson aimed his arguments at men like Morton and Agassiz who regarded "the people of color as one consolidated mass, all huddled together, without any particular or general distinctions, social or otherwise." Too many white Philadelphians assumed that the "sight of one colored man . . . is the sight of a community." To teach readers to see more clearly, Willson described a community of educated householders who lived intelligent and refined lives, furnished their parlors, and organized churches, schools, debating societies, and literary associations.[64]

Middle-class white Americans like Morton would have recognized the refined world Willson described, but white men were also to blame for putting limits on that world. Sharp anger slips into his descriptions of the "exceedingly illiberal, unjust and oppressive prejudices of the great mass of the white community, overshadowing every moment of their existence, is enough to crush—effectually crush and keep down—any people. It meets them at almost every step without their domiciles, and not unfrequently follows even *there*. No private enterprise of any moment,—no public movement of consequence for the general good,—can they undertake, but forth steps the relentless monster to blight it in the germ." He expressed a particular frustration with scientists. A man "may indeed reach the base of the hill of science; but what does he behold? Brethren ready to extend the hand of greeting and congratulation, when he hath made ascent? No; not so with the man of color: need I say what reception *he* would be most likely to meet with?"[65]

One wonders if Willson imagined Morton's features on the faces of those offering such a cold reception to dark-skinned young men who wanted to be scientists. Willson finished his book in June 1841 and dropped the pages with printers S. E. Merrihew and Lewis Thompson at their shop at 7 Carter's Alley. The firm produced a lot of antislavery material in the 1840s, but Merrihew and Thompson also worked for Morton, printing his first catalog of skulls, as well as his lectures on ancient Peruvians and his polygenist paper "Diversity of the Human Species."[66]

Even if Morton tried to ignore the abolitionist activity in his city, he could hardly have missed the reform-minded men and women, black and white, who frequented the print shop, a hive of abolitionist activity. He and Agassiz dreamed of a science that transcended current events and imagined their work immune to shifts in contemporary tastes and politics. If Morton had thought to talk to Willson as they stood waiting to check page proofs, chatting with printers, he might have learned what people thought about the connections between the work of naturalists and the country's racial politics. But racial politics set limits on Morton's friendships too. With his mind on a Phoenician skull or a flattened head or on his debts to printers and artists, stoop-shouldered Morton passed Willson in the printers' doorway and never gave the young man another thought. Of course, this is how social power works: Willson knew Morton's world; Morton could not be bothered to learn about Willson's. He was sure he knew all he needed to know.

In the 1850s, a little over a decade after Willson published his book, Frederick Douglass took up the challenge to men of science, exposing the hypocrisy behind the pose of disinterest and laying out the consequences of practices like skull collecting. On July 12, 1854, Douglass delivered a commencement address to the "Gentlemen of the Philozetian Society," one of the literary societies of Western Reserve College in Hudson, Ohio. (Case Western Reserve University today.) Morton had died in 1851, but his ideas were seeping into American life. Douglass took them on. Douglass was already famous, an outspoken abolitionist, an editor and author, who had published his autobiography, *Narrative of the Life* in 1845. He had been reading the ethnological speculations published by Morton and his friends, and that afternoon he reworked a speech he had written called "The Races of Man."

Three thousand people crowded into a tent to hear a black man speak for the first time at the graduation ceremony of a white college. Douglass began his address by remarking on the novelty of the situation—a man with no formal education invited to address college graduates. The occasion was a new chapter in his "humble experience." Douglass dropped humility as he began to assess "the claims of the Negro, ethnologically considered." The topic took him straight into the intellectual territory of Morton, Gliddon, and Agassiz, into a "moral battlefield," where the "neutral scholar is an ignoble man." The position of scientific neutrality that Morton and Agassiz pretended to hold was untenable, Douglass said. "Here, a man must be hot, or be accounted cold, or perchance something worse than hot or cold. The lukewarm and the cowardly, will be rejected by earnest men on either side of the controversy. The cunning man who avoids it, to gain favor of both parties, will be rewarded with scorn; and the timid man who shrinks from it, for fear of offending either party will be despised."[67]

Scorn would be slow in coming to Morton and Agassiz, but Douglass set scorn to work in his talk to young scholars on that hot Ohio afternoon. He had three main points to make. First, he dismissed as "scientific moonshine" arguments that questioned the manhood of the Negro. Second, he wondered that "learned men—speaking in the name of *science*" could question the "unity of the human race." "The brotherhood of man—the reciprocal duties of all to each, and of each to all, are too plainly taught in the Bible to admit of cavil. The credit of the Bible is at stake." Morton, Agassiz, and coworkers had divided men to suit their own narrow prejudices, following the misbegotten ideas that turned them against the "grand affirmation of the unity of the human race." A few ethnologists, James Cowles Prichard and George Combe, bucked the wicked trend, Douglass thought, and, in recognizing the "mental endowments of the negro," saw the true outlines of the human family. They helped Douglass visualize the multi-racial promise of America.

Reassertion of monogenesis was the heart of the speech, and it led Douglass to a third claim. Douglass shared Lowell's worries about theology and science, but reconciled the two under the authority of the Bible in order to upend the ideas behind polygenism and counter their vicious consequences. Douglass dismissed as absurd the claim that the white race could take credit for the accomplishments of ancient Egyptians. Prejudice blinded scientists like Morton and Gliddon who had "sacrificed what is true to what is popular." The sacrifice of truth, Douglass told the graduates, was not the proper vocation of the scholar.[68]

Douglass challenged Morton's logic, his science, his faith, and his politics. And indirectly, he challenged what it meant to fill a study with human skulls. As he worked through his arguments for the unity of the human race, he wandered back to the 1680s, to Reverend Morgan Godwyn's writings urging missionaries in the West Indies to baptize Africans and Indians. "The West Indies have made progress since that time.—God's emancipating angel has broken the fetters of slavery in those islands, and the praises of the Almighty are now sung by the sable lips of eight hundred thousand freemen, before deemed fit only for slaves, and to whom even baptismal and burial rights were denied," Douglass said.

Baptism and burial were basic markers of humanity for Douglass—a welcome into the human community and a communal farewell. Morton's conclusions were unsound and wicked, Douglass knew, but a quick glance at the thousand skulls in his collection made clear a powerful insult at the very start of his inquiry. A lawyer might have steered Douglass to the long history of the common law of the right of sepulcher, which gives a next of kin the right to dispose of a body. Douglass argues that to deny the right to bury the dead erodes humanity. And as he told the scholars near the end of his two hours of talk, "God has no children whose rights may be safely trampled on."[69]

Douglass met Morton on his own ground—and took a stand with his newly minted Ohio scholars on the field where biology and culture crossed. Morton's biology of race was wrong, but Douglass recognized that people took Morton's arguments seriously because he got the culture right. Morton mixed national aspiration with an obsession with race, fine lithographs with numerical tables, dead Indians with popular fascination with Egypt, sentimental friendship with hard-headed science. His skull work also appealed to a generation obsessed with death.

Over the next several decades, skull work would continue the slow erosion of burial rights, although craniologists never fully succeeded in washing off humanity and human mortality from their specimen skulls. In the last years of the twentieth century, advocates representing the collected dead pushed institutions to give up old skulls, and many of the human crania in Morton's collection have gone back into the ground. But more than a hundred years of wars and atrocities passed before cranial science gave up its dead.

(**292**) Cranium M. æt. C.40, Cap. 1495 c.c., L. 187 mm., B. 141 mm., H. 136 mm., I.f. m. 49, L. a. 391 mm., C. 467 mm., Z.d. 145 mm., F. a. 70°. "Vendovi," chief of one of the Fiji Islands. Received in exchange from the Smithsonian Institution.

—GEORGE A. OTIS, *List of the Specimens in the Anatomical Section of the United States Army Medical Museum* (1876)

Dried sea cucumbers, if not purchased presoaked, must be treated by a fairly involved process. First soak them for 4 hours in cold water and then scrub them with a brush. Place them again in cold water, bring to boil, cook 5 minutes, and allow to cool in water again. Repeat this process 10 times. The meat then is swollen and soft and ready to use.

—CALVIN W. SCHWABE, *Unmentionable Cuisine* (1979)

CHAPTER 4

"News from the Feegees"

"Our Feejee Cheif"

Friday, June 10, 1842. American naturalist Charles Pickering sat on board the sloop of war *Vincennes* in New York Harbor, writing to his old friend Samuel George Morton. Pickering had not seen Morton for almost four years, but he urged him now to hurry up to New York to meet a man from Fiji. "Vendovi," Pickering called him, was fading fast. "Our Feejee Cheif [*sic*] is on his last legs and will probably give up the ghost tomorrow." "As you go in for *Anthropology*, it would be well worth your while to come on immediately, for such a specimen of humanity you have never seen, and the probability is, that you may never have the opportunity again." As far as Pickering knew, only one other man from Fiji—a dwarf paraded through the streets of Salem and exhibited in Baltimore in the 1830s—had ever come to the States. Certainly no one from Fiji had come to the United States to become a specimen for curious anatomists or left a skull for craniologists. Pickering's sentence gives us a good use of the word "specimen" in two senses. The naturalist was in the business of collecting specimens to preserve as typical examples of the things he studied. Writing to Morton, he used "specimen" ironically, to underline the Fijian's strangeness.[1]

Only miraculous speed could have gotten slow-moving Morton to New York in time to meet this Veidovi, to adopt a spelling closer to Fijian pronunciation. Veidovi died of tuberculosis in the hospital at the Brooklyn Navy Yard on Saturday, June 12, 1842. Hospital staff buried his body in Brooklyn, but someone pickled his head and sent it to Washington. Veidovi was probably in his mid- to late thirties, about Pickering's age, when he died. There was nothing remarkable about a relatively young man dying from consumption in Brooklyn in 1842; tuberculosis killed thousands of New Yorkers every year. The real question is not what killed Veidovi, but how he came to die in Brooklyn in 1842. Life changed in Fiji while Veidovi was growing up in the first decades of the nineteenth century. Traders, missionaries, explorers, and naturalists, like Pickering, pulled Pacific islands into relationships that altered their ecology, economy, and culture. It is impossible to recover what Veidovi thought about his long trip from Fiji to New York, but we can try to understand what Pickering had in mind as he imagined Veidovi as a "specimen of humanity."

Morton had a cabinet filled with specimens of humanity, but none so rare as this living man. The craniologist had probably never seen a skull from Fiji. Even in the 1830s, skulls from the Pacific were hard to come by in Europe and America. Sailors hated carrying human remains in the holds of ships. Dead bodies went overboard. Scarcity of specimens left craniologists and ethnologists or race scientists, like Pickering and Morton, struggling to fit Pacific islanders on their racial maps. Those members of the Malay race that Morton labeled the "Polynesian Family" blurred racial borders. As Harvard anatomist John Collins Warren wrote, "It is obvious that no distinct boundary lines can be drawn between the different varieties of the species, since they are on each side passing by imperceptible grades into each other." The people of Fiji were particularly puzzling: their faces looked Polynesian, but they had wild and frizzy hair like Negroes.[2]

Pickering had written to Morton from Fiji, speculating about the people he met. If Morton had gone to New York, the two men could have compared notes on Veidovi and his body, and Morton could have heard Pickering's fresh thoughts on a fantastic voyage. While Morton minded his skulls in Philadelphia and nursed *Crania Americana* into the sagging book market, Pickering sailed around the world, one of nine men chosen for the "corps of scientific gentlemen" to accompany the United States Exploring Expedition—an ambitious "voyage of discovery" sponsored by the U.S. government and the last of the world's great naval expeditions under sail.

On August 18, 1838, six American vessels (the sloops of war *Vincennes* and *Peacock*, the ship *Relief*, the brig *Porpoise*, and the tenders *Sea-Gull* and *Flying-Fish*) headed off from Norfolk, Virginia, under the command of Lieutenant Charles Wilkes, to explore and survey the "Southern Ocean." The expedition (the "U.S. Ex. Ex.," in contemporary shorthand) visited ports along South America's Atlantic coast, rounded Cape Horn, sailed far enough south to prove Antarctica a continent, charted Pacific islands (including the little-known Fijian archipelago), and mapped the Oregon coast before heading west across the Pacific, through the Indian Ocean, around the Cape of Good Hope, across the Atlantic, and back to New York. Congress provided the expedition with a "liberal appropriation," an investment to help support American commerce, particularly the substantial industry of commercial whaling. Estimates put the value of the whale fleet at about $25 million (roughly half a billion in today's dollars). American whaling produced about 400,000 barrels of oil, "yielding an annual return of five millions, extracted from the ocean by hard toil, exposure, and danger."[3]

To help reduce that danger, the expedition ventured to the "doubtful islands and shoals" of the Pacific. The U.S. Ex. Ex. promised to return with the maps and charts that would let sailors predict the winds and avoid hidden reefs. In the short term, any new knowledge about the Pacific would help whale men, but for those who took a longer view, charts of the Pacific were a means to secure the United States a place among the nineteenth century's small group of great maritime powers. The young country's confidence in its future was astounding. When the expedition left Virginia in 1838, the United States had yet to claim land on the Pacific, although fur traders (John Jacob Astor) and missionaries (Jason Lee and his Methodist friends) had already set their sights on Oregon. Many Americans believed in their nation's future on the Pacific, subscribing to a national destiny that reached that ocean and then crossed it to touch Asia's fabulous wealth.

Lowly facts fed these lofty dreams. The U.S. Navy Department's official instructions to the U.S. Ex. Ex. emphasized practical knowledge of currents, winds, and ports. But the expedition aspired to more. Pickering and his colleagues in the corps of scientific gentlemen joined the U.S. Ex. Ex. "to extend the bounds of science, and promote the acquisition of knowledge," whose immediate applications were less clear.

Hydrography, geography, cartography, philology, botany, horticulture, conchology, mineralogy, and ethnography all contributed to a grand American future. The scientific gentlemen fished for facts and specimens, sometimes with a drag net that turned up the marvelous and unexpected, and the unwilling.[4] Veidovi was an unlikely and unwilling specimen caught in the search for knowledge.

In the early summer of 1842, curiosity sparked around Veidovi and his body. New Yorkers, delighting in the odd visitor, chattered about the "sailor eating" "Fegee chief." Think of the tales he could have told about Fiji, a place Americans tagged with unhappy derogatory superlatives: home to "the most barbarous and savage race now existing upon the globe." According to Morton, Fijians vied "with New Zealanders in treachery and cannibalism." Like other Polynesians, they were men of "volatile disposition and fugitive habits. They act from the impulse of the moment, without reflection and almost without motive. Thus they are kind or cruel, loquacious or taciturn, active or indolent, according to the promptings of caprice or passion; and they have been truly said to possess the foibles of children with the vices of men."[5]

Sailors described Fiji as a labyrinth of treacherous coral reefs. Experiences in Fiji defied European and American fantasies that Pacific islanders were all there to welcome sailors home from the sea. Not Fiji. Fiji was not the romantic Pacific of scantily clad dancing women scampering in the surf on sun-kissed beaches. Dark mangrove swamps lined Fiji shores, and those thick mangroves sheltered armed savages, just waiting for boats filled with innocent sailors to ground on the coral reefs. By Fiji custom, sailors said, Fijians had the rights to take any grounded craft and anyone or anything a grounded craft contained. In the early years of the nineteenth century, those grounded crafts contained Europeans. A fortunate few fell under the protection of local chiefs and made themselves important by repairing rusty muskets. Fijians roasted and ate the less fortunate. Or so sailors said. Reports of hidden reefs and stories of hungry cannibals pushed sailors headed for Fiji to write out their wills. Reports worried American traders, too, who cultivated markets in Manila and Canton for sandalwood, tortoiseshell, and bêche-de-mer but had trouble finding insurance companies willing to underwrite voyages bound for Fiji.[6]

Veidovi didn't have a chance to tell his story. Or we have no record that he did. His quick death left curious Americans without answers to questions they would have asked him—whether about Fiji's reefs, Fiji politics, or Fijian appetites. Pickering and others recorded fragments of conversations with him. Missionary reports, chronicling the wars and rivalries that roiled Fiji in the 1840s and 1850s, provide some context for Veidovi's experiences. We have a better sense of the physical dimensions of the man. Craniologists measured his skull. Some hoped his corpse could be made to tell a story the man could not. There were rumors that Dr. Valentine Mott, one of New York's most important surgeons, would invite the public to watch him dissect Veidovi. There is no record that Dr. Mott cut up the corpse for gawking New Yorkers, but someone did cut off Veidovi's head, and it is likely that those who took his head also took a look at his diseased lungs. Someone then buried his headless corpse in the cemetery at the Brooklyn Navy Yard, although records of the grave's exact location have been lost in the Navy Yard's moves and expansions in the years since 1842.[7]

It has been easier to keep track of his skull. Or somewhat easier. The day after Veidovi died, the *New York Herald* reported that "surgeons at the Hospital have already cut off his head, and it has been laying in pickle for several days, before the process of embalming takes place." The story continued in a New York vein, "The hair of his head is thirty inches long on each side, and in his days of glory, he had thirty hair dressers to wait on him. Vendovi beat the Broadway dandies all to smash." Expedition naturalist Titian Ramsay Peale (1799–1885)—a member of America's first family of artist/scientist/collectors and Pickering's colleague in the expedition corps of scientific gentlemen—cataloged "Vendovi's cranium" as item number 30 in a handwritten list he compiled of the expedition's collection of some four thousand ethnographic objects.[8]

New Yorkers jumped at Veidovi's story, incorporating it into a lively popular culture of penny newspapers and dime museums. Sunday's *Herald* shouted, "LAST OF VENDOVI!—The Fegee Chief Vendovi, who arrived here in the Vincennes, died yesterday morning. He was a cannibal of the 'first class,' and has been out of health since his capture, in consequence of having nothing but roast beef and salt pork to eat. It will be recollected that several years ago, he captured a Salem brig, the Charles Baggett, and took eleven American sailors as prisoners. These he called 'tit bits,' which his Fegee Highness and family ate for dinner. It is a pity, in one sense of the word, that Vendovi is dead. He was the only real curiosity brought home by the Exploring Squadron." A day later the paper wondered about his corpse, giving the whole story a happy creepiness. "What in the name of mystery has become of the body of the great Cannibal Chief—the sailor eating Vendovi? Where is his head? A great many ladies are exceedingly anxious to know. And numbers desire that it may be placed in some Museum where they may be able to see it. There exists a strange and morbid curiosity to see the remains of this miserable wretch."[9]

The morbid curiosity that excited New York ladies failed to rouse Morton. Perhaps the craniologist sensed that the expedition's horde of ethnographic treasures—its hundreds of war clubs, yards of bark cloth, and crates of clay pots from Fiji; baskets and feathered headdresses from California; fishhooks, rattles, drums, and a few human skulls—signaled an end to the amateur scientific establishment he knew so well. One historian described the expedition materials as simply outrunning "the intellectual resources of the country." Had Morton rushed to New York hoping to meet the dying Fijian, he might have met his own scientific world on its last legs. Veidovi's skull—cleaned, measured, and numbered—became a specimen in a system the Fijian could not have imagined, but that specimen also helped usher in the new world of statistical sampling, government science, and large museums difficult to imagine for Morton's generation.[10]

Intellectual foundations had begun to change too. Pickering clung to Morton's notion that human races represented different species and that species did not change. In conversation, Morton might have sensed Pickering's intellectual frustration as he tried to accommodate experiences among the people of the Pacific to a world mapped into fixed and unchanging racial categories. Pickering tried an almost mind-boggling multiplication of racial categories. Blumenbach had said five. How about eleven or thirteen or maybe more? It would be another decade before Charles Darwin voiced his conviction that species were not nature's immutable categories, but he discovered the germ of that insight aboard the *Beagle* in the 1830s. Slowly, it changed the conventional wisdom that shaped Morton's intellectual world.

The man from Fiji draws us back into this complicated world. Like every man or woman whose remains were collected and displayed or every man or woman whose remains have been lost to a community of mourners, Veidovi left a history in fragments. It is one of the marvels of the archives that we can find fragments and begin to piece his story together again. Veidovi brought "news from the Feegees" to the United States, a sure sign that life was changing on those islands. His story moved into the lively contradictory world of 1840s America, a culture that mixed high-minded curiosity with loud laughter, brave aspiration with quick violence, and, at its base, freedom with slavery. But his story also illustrates how America's imperial aspirations helped to invent Fiji as a place of small islands, launching descriptions that emphasized the archipelago's relatively small landmasses rather than the vast seas that linked the islands. As island scholars have pointed out, this was a continental perspective.[11]

Hints to Craniographers

Morton missed his chance to meet Veidovi, but he and his intellectual heirs encountered the man's remains. Morton may have seen Veidovi's skull when some of the expedition materials on their way to Washington were temporarily placed in his care in the early spring of 1851, but he did not leave his opinion of this skull. After Morton's death that May, Mrs. Morton cleaned out her husband's study and sent his skulls across town to the Academy of Natural Sciences. One of Morton's younger craniologist friends James Aitken Meigs took charge of the collection. Meigs unpacked the skulls and began to imagine using them to establish more systematic grounds for craniology. He checked some of Morton's measurements and revised the collection catalog to include skulls donated since Morton's death.[12]

Meigs played a small part in the history of American race science. He didn't have Morton's stature, Gliddon's panache, Agassiz's ambition, or Combe's warmth, but in his quiet way he carried craniology's lessons out of Morton's study. Friends remembered Meigs as a studious child, describing him as a little boy who passed up birthday toys with a happy shout, "Oh! A book, a book!" Grown-up Meigs kept up these bookish ways, practicing medicine in Philadelphia, teaching at Pennsylvania medical schools, and studying climatology and the physiology of stammering. He published new skull measurements in "The Cranial Characteristics of the Races of Men," an essay that appeared in *Indigenous Races of the Earth*, one of the volumes edited by Egyptologist George Gliddon and Alabama physician Josiah Nott that secured Morton's place in the heart of nineteenth-century scientific racism. Meigs made a case for craniology as a necessary tool for "the American citizen" "since upon American soil, representatives from nearly all parts of the earth have been gathering together during the last two hundred years." He continued Morton's quest for answers to questions about race that puzzled many Americans, whatever their racial backgrounds. Their answers were all wrong, but they asked compelling questions about the origins of racial order. Since there was no room for change or chance in the world Morton and Meigs imagined, craniology revealed the unchanging characteristics of human races. And those unchanging characteristics indicated that God had blessed a racial order that accorded the largest share of the world's power and wealth to white men.[13]

And the racial order's logic would all be clear if only craniologists could get their hands on enough heads. Meigs had a simple plan, which he published in 1858 as *Hints to craniographers*. Craniologosts needed to catalog their "cranial collections, which constitute, so to speak, the store-houses of the raw material ready to be elaborated into a science."[14] Meigs went touring those storehouses in the 1850s, checking on skulls in Philadelphia's anatomical collections, medical museums, and phrenological societies, and then heading to Washington, D.C., to look at some heads and skulls in collections of the National Institution for the Promotion of Science. Visitors called the collection "The National Cabinet of Curiosities." The Great Hall of the United States Patent Office was the only room in Washington large enough to hold the collection and accommodate the crowds who wanted to see it.

That was where Meigs found Veidovi's head, once he elbowed past cases of fossil ammonites, stuffed birds, pressed flowers, petrified wood, coral specimens (still in the Smithsonian's remarkable collection), and "fishes in alcohol, not arranged." Or pushed past displays of new technologies to produce dehydrated milk for sailors and chromolithographs for American parlors, glanced at a necklace made of human teeth (to remember a dead friend, his guidebook explained), wooden idols, feathered headdresses, and "Specimens from Egypt" (gifts from Morton's friend George Gliddon). Benjamin Franklin's printing press was in a case. There was a portrait of President John Tyler, a lock of General Simón Bolívar's hair, a silk hat worn by an organ grinder's monkey, the mummified head of an alligator, Davy Crockett's tomahawk, Andrew Jackson's "Military Coat," a bust of Robert Burns, and the white silk breeches George Washington wore when John Trumbull painted his portrait.[15]

No surprise that the sprawling mess of a collection contained discordant elements, expressions of the young country's contradictory history. Meigs saw triumphs: the "Original Declaration of Independence, Relics of General George Washington, Treaties with foreign Powers, and Presents to officers in Government employment," including gifts the emperor of Japan presented to Commodore Matthew Perry. But there were darker stories in the length of rope confiscated from an African man who had braided it from native grasses he snatched as he was herded aboard a slave ship, in a queue that taunting sailors cut from the head of a Chinese man, in the helmet and cuirass stripped from the body of General Santa Anna's bodyguard (blood-soaked trophies from the recent war in Mexico), and in the large collection of "most curious and frightful bludgeons" from "Fejee." War clubs, carved from the hard roots of mangrove trees, beautifully designed to smash a human skull.[16]

Fiji dominated the National Institution's display of ethnographic materials. Curators cataloged more than 1,200 items from Fiji, almost twice the number of objects collected from the rest of Polynesia. (There were 110 items from Micronesia and Melanesia, 12 from Australia, 300 from North America, 91 from South America, and 52 from other places.) Passing displays of war clubs and bark cloth, Meigs found the skulls he'd come to see in Case 37, which contained "a very superior collection of human crania, many of them collected by the U.S. Ex. Ex. from the Pacific Islands." There were specimens from the Columbia River and Peru, which would have been familiar to Meigs. Pacific skulls were more interesting. Morton had half a dozen skulls from the Sandwich Islands. (Ornithologist John Kirk Townsend dug them up from a cemetery on his trip home from Oregon and gave them to Morton.) He had one skull "from the Marquesas, from the village of Whytahoo, Resolution Bay, in the Island of Christina, where it was obtained in 1841, by Lt. H.A. Steele, U.S.N., for Dr. L.P. Bush, and by the latter presented to me. F.A. 82°. I.C. 90.5." "The Christina Islanders are cannibals," either Morton or Meigs noted. And he had two shrunken heads from New Zealand, souvenirs from sailors and whale men, who sometimes peddled them to curious collectors. (Picture Herman Melville's Queequeg hawking his embalmed heads along New Bedford streets "for all the airth like a string of onions."[17])

Meigs paused at a case full of specimens from the Pacific, including a "New Zealand Chief, Pekea of Taunka, of Noot-koo-koe tribe," and "Tattooed heads—one the head of the Feegee Chief, Vendovi." The description suggests that Veidovi's embalmed head (not a bare skull) was on display. If Meigs consulted his guidebook, he knew that curators thought Veidovi's head "one of the most interesting in the collection and calculated to excite feelings very different from those experienced in examining other specimens." This head represented a chief, a cannibal, and a murderer—social facts difficult to conjure from Meigs's cold calculus of skull measurements. Notice that Veidovi's "features are decidedly different, hair in remarkable good preservation; teeth in very good order, and sharp, as they are filed with the shark skin, so as to keep them ready for use." To remind visitors what those teeth had been used for, curators displayed Veidovi's head beside a head that expedition officers bought in Fiji from a man they saw munching on its eyeballs. The head with eaten eyeballs had cost the Americans a "fathom of Blue cloth." The paired heads made a nice display: Veidovi, the famous cannibal, alongside the head of a man who had been a cannibal's meal.[18]

But what did Meigs, with his dreams of objective knowledge produced by calculation and measurement, make of the display? Skull displays were a normal part of Meigs's world. But here was a case whose grisly associations disturbed dreams of a systematic scientific craniology. Did Meigs look into the allegations of murder and cannibalism that had been leveled against Veidovi? Did he wonder how this man had come to die so far from home? Did he see that history shadowed his specimens? The exhibition in the Patent Office had hints for a craniologist curious to learn Veidovi's history. If Meigs made his way back through the exhibition to the display of "fishes in alcohol," he would have found a jar labeled *Holothuridae*, containing a pale-looking specimen of bêche-de-mer (holothuria, sea slug, beach le mar, *bicho do mar* in Portuguese, *trepang* in Malay, *swalloo* in Borneo, or *dri* in Fijian). The story behind Veidovi's skull begins with these sea creatures.

Perils of the Sea Slug Trade

Early in the nineteenth century, just about the time Veidovi was born, shipwrecks, sandalwood, turtles, sea slugs, and yams drew Fiji into European and American imperial orbits. Metal tools taken in trade and muskets recovered from wrecks changed life in Fiji. So did European microbes. An epidemic of cholera or dysentery, a wasting sickness Fijians call *na lila balavu*, tore through Fiji's virgin soil, ravaging coastal populations. In places it carried a force of cultural destruction and left communities too devastated to bury the dead. Sailors carried measles, pneumonia, tuberculosis, gonorrhea and syphilis, and kegs of rum.[19] Although only a few white men imagined making lives on Fiji's lands,

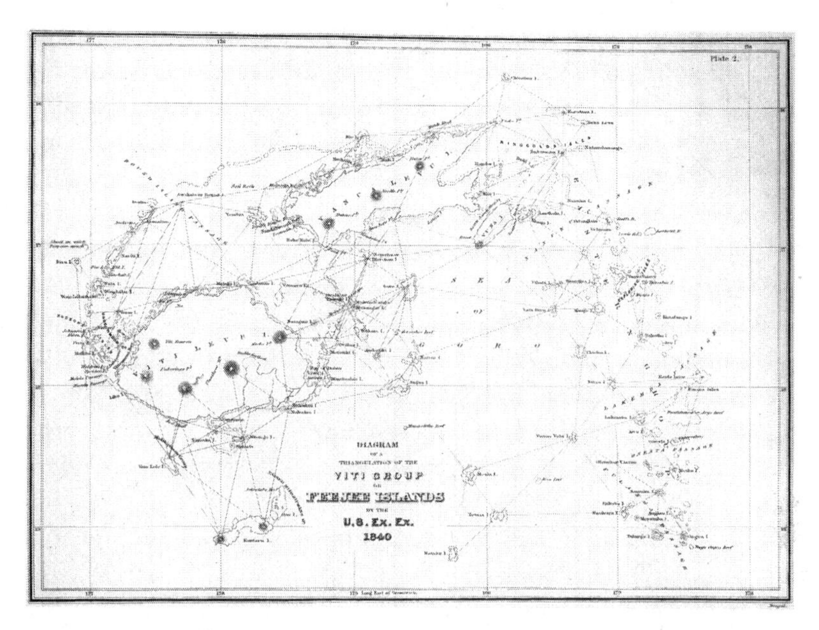

Fig. 20. Diagram of a Triangulation of the Viti Group or Feejee Islands by the U.S. Ex. Ex. (1840)
(Courtesy of the Smithsonian Institution Libraries, Washington, D.C.)

the violent confrontations and threats of extermination frequent in settler societies in Australia, Africa, and North America hit Fiji too.

The islands of Fiji spread out across 300 square miles of the South Pacific between 15°30′ and 19°30′ south latitude and 177° east longitude and 178° west longitude. Wilkes counted 154 islands scattered over the ocean, like scraps of paper blown across the floor. The largest island, Viti Levu, measures about 4,000 square miles, a bit smaller than the state of Connecticut; Vanua Levu, the next largest, measures about 2,100 square miles, a bit larger than the state of Delaware. In the nineteenth century, people lived permanently on at least sixty-five islands, visiting others to gather wood or to fish. Coral reefs surrounded many islands, making the passage through Fiji particularly treacherous for deep-drawing European and American vessels, "a perfect labyrinth," according to one American officer.[20]

When Europeans and Americans began to eye this section of the Pacific at the start of the nineteenth century, they knew very little about Fiji. Geography books credited Dutchman Abel Janszoon Tasman with the discovery of the islands in 1643. James Cook passed through Fiji on his second voyage in 1774 but did not stop. William Bligh and nineteen survivors of the mutiny on the *Bounty* sailed through in their open boat in 1789. The Russians and the French had visited, and whalers occasionally stopped to trade iron axes and metal fishhooks for wood, water, pigs, poultry, and yams. English Methodists ventured out to Fiji in the early 1830s. Fijians tolerated their presence, but very few rushed into the Christian fold. Missionaries found a few European mutineers, deserters, and shipwrecked sailors who had washed up in Fiji, made themselves useful to local chiefs, married Fijian women, and stayed put. Missionaries considered these beachcombers an unsavory lot, but they served as a wedge of white settlement, arming Fijians with muskets and working for European and American visitors as pilots and interpreters.[21]

In 1840 Wilkes estimated Fiji's population at about 130,000. Most Fijians lived on the islands of Kantavu (13,500) and Ovalau (8,000), and at Muthuata (15,000), although the real centers of power were at Rewa (population 5,000), a settlement that ran a mile up from the mouth of the Rewa River on Viti Levu, and at Bau (or Ambau—population 3,000), an island village just off the west coast of Viti Levu. Early in the century and again between 1843 and 1855, Rewa and Bau fought bloody wars to dominate Fiji. Their animosity surprised Western visitors, who described Rewa as little more than a "dirty muddy hole," especially during the rainy season. In dry months, visitors discovered a picturesque village, large chiefs' houses alongside rickety shanties for ordinary *kaisis*, high-roofed *mbures* or spirit houses, which doubled as shelters for travelers. Rewa's alluvial fields teemed with yams, breadfruit, coconuts, and bananas, which Americans thought grew with little labor. The plenty troubled sailors who objected to such easy abundance, although Rewa's fertile soil assured Europeans and Americans the cheap provisions they had come to Fiji hoping to find.[22]

Rewa was the power base for Veidovi and his three half-brothers: Ro Kania (Tui Dreketi, the king of Rewa), Ratu Qaraniqio (Ngaraningiou), and Ro Cokanauto (Thakonauto). The Americans also knew Ro Cokanauto as "Phillips," a name he adopted from the owner of a vessel that had taken him to Tahiti. Phillips spoke English, French, Spanish, and some say Italian, as well as Tahitian, Tongan, and many of Fiji's dialects. He fascinated visitors with his bits of European furniture (cups and saucers displayed on a buffet, mirrors hung on thatched walls, frying pans, teakettles, and a music box that mimicked birdsongs all arranged to give his Fijian home the look of an English farmhouse), his pieces of European clothing, and his conversation: "All hands of us bothered him with questions that we could not resist asking, for to us it was the very height of novelty to converse with a Fegee in [our] own tongue," Midshipman William Reynolds remembered.[23]

Veidovi would have been Ro Veidovi to these Fijians, a member of the ruling family of Rewa. By his mother, Veidovi was related to those who ruled at Kantavu, which gave him a nephew's or *vasu*'s privileges there. Anthropologists understand the *vasu* as part of a complex kin system that provides an underlying structure to relations among Fiji's towns and villages and helps explain Fiji's rivalries and wars. Americans thought *vasus* were just bullies who demanded tribute and intimidated their relatives' villages. It surprised Wilkes that towns sometimes received a *vasu* with honors greater than those paid to a chief. Bowing and clapping, the people shouted, "Hail! Good is the coming hither of our noble Lord Nephew." Wilkes thought it all an empty performance, and when he sailed off with Veidovi on board, he described the people of Kantavu as happy to be rid of troublesome *vasu* Veidovi.[24]

It is possible to tease out information about the history of Fiji in American and European accounts, although many Western chroniclers of nineteenth-century Fiji move through a repertoire of set pieces on Fiji and Fijian culture, as if a tourist were checking off sites on an organized itinerary: the burial of a chief and the sacrifice of his wives, the mutilation of mourners, the intentional killing of aged relatives, stylish dances, elaborate hair, quick tempers, great cunning, constant treachery, and then, always just offstage, feasts on human flesh. Rarely do these things appear as specific events. They are the timeless aspects of Fijian culture, in Wilkes's phrase, the "customs of the Feejee Group."[25]

In American eyes, the coming of traders and missionaries helped bring history and events to Fiji, although Westerners on extended stays (missionaries, beachcombers, and sea slug traders) learned that Fiji's history was as complex as that of any European country. Struck by a country of small tribal communities, divided by rank, ruled by rival chiefs, and torn by constant warfare, Western observers turned to their own history. Expedition philologist Horatio Hale compared Fiji's many chiefdoms to the warring republics of ancient Greece and described island intrigues that suited a Machiavellian principality. A mid-century British Methodist thought Fiji shared France's gift for war and Italy's taste for religion.

> Here a Fijian Rome and a Fijian Carthage have had their Punic war. Here exist their maritime nations, answered to the flourishing empires of the European world; and here are the Fijian Spain and Portugal, without a navy. However trifling Fijian history may appear to a stranger, its details grow in importance to those who live amongst them; and the student who has watched them in the place of their occurrence may surely be excused if he has sometimes allowed himself to be amused and interested by thus tracing historic parallels, where the principles involved and the motives at work were the same, and where the diversity in their operation and results was caused only by the difference of magnitude in the forces employed and the field on which they moved.[26]

Outsiders began to appreciate the intricacies of Fiji's history in the first years of the nineteenth century when American and European traders saw wealth in the sandalwood trees that grew on the rocky slopes of coastal Vanua Levu. The oldest wood in the butt of the sandalwood contained oil that Polynesians used to perfume the coconut salves they rubbed on their skin. Tongans traded bark cloth, sail mats, and the barbs of the stingray for Fiji's sandalwood. Once the whalers reached the Pacific, Tongans bargained for wood with nails, axes, chisels, and whales' teeth. But demand for sandalwood and its oil in India and China changed the balance of the trade between Tonga and Fiji. In China, carvers turned sandalwood into ceremonial artifacts. They pressed sawdust into joss sticks and burned it as incense.

By 1809 much of Fiji's sandalwood had gone up in smoky perfume in temples in India and China. Storms at sea and violence on shore made the trade risky, but profits were good, producing returns as high as 600 percent for bold traders who changed sandalwood into tea, coffee, sugar, and silk in markets at Manila and Batavia. Sandalwood networks drew Fiji's people into permanent relations with America and the West. Fijians began to exchange their labor for iron tools, tobacco, muskets, cloth, and whales' teeth, which changed the ways they planted and fished and calculated political power.[27]

When the sandalwood was gone, Fijians and traders searched for other means to maintain the profitable traffic in and out of Fiji. A few looked to tortoiseshell, but "bêche-de-mer"—as Westerners called the dozen species of Holothuroidea that thrived in warm water on Fiji's shallow reefs—promised easier profits. The first American bêche-de-mer traders, half a dozen Salem captains, went to work in Fiji in the 1810s. They recruited Fijians to collect fish on the reefs, but before sea slugs could be shipped to markets in Manila or Canton, the watery creatures had to be cleaned and dried. At first traders spread cleaned sea slugs in the open air, hoping rain would hold off long enough for the fish to dry, an uncertain prospect in Fiji's fickle climate. To reduce uncertainties, traders built drying houses (constructed of coconut logs and measuring about 85 feet by 20 feet) and hired men to tend the sea slugs spread on racks over smoldering mangrove logs. Americans officers sat in an "elevated trade stand" to supervise the Fijians, Tongans, and Hawaiians, who collected, cleaned, and dried the fish, and to guard the "articles with which the fish is purchased."[28]

This more efficient processing quickened the trade's ecological and cultural consequences. Fish and forests disappeared. Fijians, who worked for traders, dropped traditional pursuits of warfare and agriculture. Bêche-de-mer made fortunes for a few. Dried fish were easy to ship, since the fish "were hard, of the consistency of sole leather," and would stack up easily in the hold of a ship. (An unlucky flood in the hold turned dried fish into a glutinous mass.) Most American traders sold bêche-de-mer to merchants in Manila, who sent it on to Canton, where Chinese cooks, who prized the fish as an aphrodisiac, made it into soup.[29] (Though bêche-de-mer smells fishy, feels like jelly, looks like a large leech or bloodsucker, and has very little taste, Chinese cooks believed it improved "a man's virility and sexual capabilities," according to a food historian. Perhaps the shape of a living sea slug or one inflated after a good soaking—eight inches long, three inches thick—was just suggestive enough to do the trick.) Chinese appetite created a strong market.[30]

A cargo of bêche-de-mer weighed about 1,200 *piculs*, a Malay word for a traditional Chinese measurement equivalent to about 133-1/3 pounds, thought to be the amount a grown man could carry on his back. Since each *picul* of dried fish consumed about half a cord of firewood, 600 cords of mangrove logs, cut and burned, went into each shipment. But those 1,200 *piculs* of fish, which left bare acres behind in Fiji and promised great happiness in Chinese beds, could produce as much as $27,000 from traders posted at the Manila market. American traders calculated that a supply of muskets, fishhooks, knives, scissors, chisels, gouges, and whales' teeth needed to purchase labor in Fiji cost only $3,500. For Americans, the trade was lucrative, easily worth the risk of the occasional violent encounter.[31]

Fijians entered into the new calculus too. During the late 1820s and early 1830s, bêche-de-mer traders flooded Fiji with small iron tools, muskets, and whales' teeth, changing labor practices, altering warfare, and inflating the symbolic currency of power. Beachcombers liked to brag about the changes they brought to Fiji when they introduced muskets, but Fijians actually incorporated new weapons and new tools into their old practices. Some Fijians were reluctant to join the fishing enterprise, preferring to see themselves as warriors rather than workers, but the sea slug trade played into Fijian warfare. A man who cut twenty cords of wood or caught fifteen hogsheads of bêche-de-mer could trade for a musket. He could trade one hogshead of bêche-de-mer for a whale's tooth. Both guns and whales' teeth figured in the ways Fijians wielded power, and the bêche-de-mer trade introduced more of both.

Some witnesses noted the role of whales' teeth (*tabua*) in Fiji's symbolic economy. Even as the bêche-de-mer pulled Fiji into the nineteenth century's global markets, Fijians used whales' teeth to placate angry gods, to clear a soul's path to the afterlife, to cement alliances, to make marriages, and to settle personal disputes. "A whale's tooth is about the price of a human life," Wilkes wrote, describing a case in which an offering of a whale's tooth ended a long blood feud's cycle of killing and revenge.[32]

Bêche-de-mer stayed plentiful in Fiji through the 1820s, but by the early 1830s efficient trading practices and Chinese appetites (or Chinese lust) had destroyed fish populations. Holothuria had nearly disappeared from the reefs to the west and north of Vanua Levu and to the southeast of Vitu Levu. Contemporaries recognized that this meant the end of the trade: once fishermen cleared bêche-de-mer from a reef, a sailor wrote, "it is a long time before it is again covered, perhaps never." When the U.S. Ex. Ex. arrived in Fiji in May 1840, Captain J. H. Eagleston (the man who had brought a Fijian to Massachusetts in 1835) was the last of the American traders still working the islands.[33]

After a few months in Fiji, naturalist Pickering puzzled over the bêche-de-mer trade: "To look at the small size of the prepared animal, reflect on the immense amount of industry there expended in making up a cargo, the complicated process of its preparation, &c., &c.; and all carried on by means of such people. . . . It is certainly one of the most remarkable branches of business in the Annals of Commerce."[34]

Tensions in Fiji mounted as fish supplies declined in the early 1830s, and disputes broke out between American traders and between Americans and Fijians. Veidovi was involved in one of these disputes, although the precise story behind the fight at Kantavu (a town that owed allegiance to Veidovi as *vasu*) between sailors working for Captain George Bachelor of the brig *Charles Doggett* and a small group of Fijians is impossible to recover. In New York in 1842, the *Herald* treated the story as a joke, describing Veidovi making "tit bits" of men he had killed. Americans closer to the scene thought greed led Fijians to kill the sailors—Americans, Hawaiians, and Tongans working for Captain Bachelor. Dwindling bêche-de-mer supplies and Fijian politics were also behind the fight.[35]

Ro Veidovi, Veidovi, Veindovi, Vendoba, Vindovi, Vendovi

The U.S. Ex. Ex. arrived in Fiji in May 1840. Within a few days, Wilkes heard the accusations against Veidovi from a beachcomber named Paddy Connel (or O'Connell), a seventy-five-year-old Irishman, sporting a long gray beard and nothing but a "dirty rag about his loins." Wilkes knew that eight Western vessels, five of them American, had wrecked or lost crewmen in the Fijis since 1828 and that anger over those losses helped secure support for the expedition, yet he wrote that Connel's report was the first he had heard of the murder of the crew of the *Charles Doggett*.[36]

But prior knowledge of losses helps explain why Wilkes was so ready to act on information from a source described as a "most remarkable character . . . whose history, could it be written, would far exceed in interest & wonderful adventure, the imaginary life of Crusoe, or any other tale of the marvelous, true or false." Connel said he was born in county Cavan but left home to enlist in Napoleon's fight against their English enemy. The English captured Connel, charged him with treason, and shipped him off to the penal colony at Botany Bay. Once he arrived, no one could find the charges against the transported man, and he boarded a French vessel bound back to Europe. "When they got to Sea, they did not like the way things went on board her & suspected she was a Pirate. So in passing between the Fegee Islands, 30 of the crew left the Ship with their arms to try their fortunes on shore," where they joined the community of European shipwrecks and deserters that had grown up at Levuka on the island of Ovalau.

Connel and his comrades (including a steadier New Hampshire man David Whippy, who became Wilkes's particular informant) took part in Fiji's intrigues, repairing muskets and siding with warring factions. They also entertained passing crews, like the men of the U.S. Ex. Ex., with tales celebrating their own marvelous exploits, their many wives, and their close encounters with cannibalism.[37] Connel, whose figures were never precise, said he had married 150 Fijian women, but fathered only 45 "*childer*." He brought them up not to eat human flesh, he reassured his American audience. He would die a happy man, he said, if he could round out his progeny and father just five more. (In April 1841 Wilkes received a letter reporting that Connel died disappointed "without having any more issue.")[38]

Connel's sprawling yarn contained a nugget of Veidovi's story. Connel told Wilkes he had joined Captain Bachelor at Rewa in August 1833 to help him negotiate a "contract with Vendovi, a chief of that island, and Vasu of Kantavu," to organize his people to fish for bêche-de-mer. According to Connel, Veidovi caught a glimpse of Bachelor's supply of trade goods—muskets, whales' teeth, and cloth—and decided to rob him instead of putting his people to work. At Kantavu, Bachelor took a man a hostage to insure the good behavior of Fijian workers, but when the hostage pretended he was sick, Bachelor released him. The hostage release signaled plotters, who set fire to the bêche-de-mer house and clubbed the workers who ran from the burning building. The mate and seven, eight, or nine men were killed. (Versions differ.) One man escaped.[39] When the fighting stopped, Connel went to bargain "with the natives for the bodies, seven of which were brought down to the shore much mutilated, in consideration of a musket." He said that he and Bachelor tried to bury the bodies at sea, but they "floated again, and fell into the hands of the savages," who cooked and ate them, even though "they did not like them, and particularly the negro, whose flesh they said tasted strong of tobacco."[40]

Flesh with tobacco taste carries a hint of Connel the yarn spinner, a man many suspected of "*romancing* a little." Wilkes chose to believe him since "there were no witnesses against him, & we could not separate the truth from the fable."[41] It is not clear why Connel decided to accuse Veidovi of murder or why Wilkes decided to act on his story. Connel's accusations helped him appear important to the visiting Americans, but he may have been plotting his place in Fiji too. Perhaps he had taken sides in disputes between Veidovi and his brothers or disputes between Rewa and Bau and was simply following Fijian orders when he came tattling to Wilkes. Wilkes believed Connel's account and felt that Veidovi's capture would teach Fijians not to molest American traders. Wilkes paid Connel a musket to carry a letter to William Hudson, captain of the expedition's *Peacock*, which was already anchored at Rewa. "It is earnestly to be desired that some steps should be taken to obtain the chief, or perhaps destroy the town where the offence occurred," Wilkes wrote. "Your own judgment will, however, be your best guide as to the course to be pursued; that which you adopt will be satisfactory to me." Orders in hand, Hudson set a trap to catch a Fiji chief.

As a first step, Hudson went to see Veidovi's brother, Tui Dreketi, a man he called the king of Rewa. The two had met earlier when Tui Dekreti signed the trade agreement, which included the clause that "murderers of foreigners [were] to be given up, or punished on shore." The king received "us kindly," Hudson wrote. Men and women smoked cigars. Tui Dreketi shared his *kava*, the mildly intoxicating ceremonial drink Fijians brewed from the root of a pepper plant, *Piper methysticum*. Fijians danced and Americans fired off blue rockets. The meeting was friendly, though Hudson thought hospitality had gone too far when a young woman climbed under his mosquito netting and snuggled into bed with him. Chaste Hudson slipped outside. Hudson invited Tui Dreketi and his household to visit the *Peacock* the next day, promising presents for everyone. He knew Fijians liked presents of cloth, vermilion, and beads, and liked to tour American ships. "I had every reason to believe from what I learned at Rewa—that the Chief *Bendovie* would visit the Ship in the morning for his present—and this [would] afford me an opportunity of making him a prisoner—without proceeding to other—and more unpleasant measures," Hudson wrote.

The next morning Tui Dreketi, a favorite wife, his brothers Qaraniqio and Phillips, and the "usual retinue" of fifty or sixty people climbed aboard the *Peacock*. Veidovi was not with them. Hoping he would arrive during the day, Hudson entertained his visitors, showing off his ship. Late in the afternoon, when there was still no sign of Veidovi, Hudson piped everyone "to quarters," turning his guests into hostages. Someone, probably Connel making himself useful, explained to Tui Dreketi that Hudson "had to perform the painful and unpleasant duty of retaining them all on board the Ship as prisoners until the chief Bendova should be brought on board." The people muttered, warning the Americans that relatives in Rewa would be angry if Hudson took Veidovi. They might even attack the two British Methodist families, Christianity's tiny toehold at Rewa. Hudson threatened to shoot anyone who tried to leave the *Peacock*.[42]

(Nothing happened that day. Years later a Christian convert from the Friendly Islands remembered the uproar when the news of the royal captivity reached Rewa. "The warriors ran together, vowing to kill us all because they thought the missionaries had a hand in the deed." A chief calmed the crowd, "so we escaped; but indeed we thought our time had come." British captain Edward Belcher complained that Fijians refused to trade with him "in consequence of this affair." "Indeed, the abstraction of the Rewa chief by the Americans has irritated the natives amazingly, and will probably injure mercantile interests.")[43]

Hudson ignored the threats and with Connel's help persuaded the king to send Qaraniqio to bring their brother Veidovi to the Americans, "alive if possible." (Or as Midshipman Reynolds remembered, "Alive if he could, but to kill him and bring his body if he resisted.") Americans could have had only a vague idea of events in Fiji's recent past, although Connel may have described the rivalry between Qaraniqio and Veidovi and encouraged Hudson to manipulate their enmity. Hudson congratulated himself: "The temptation to get rid of so powerful an adversary was an opportunity not to be lost by a Feejee man."[44]

Qaraniqio sailed off, leaving Hudson with a boatload of unhappy guest/hostages. Hudson invited the king and queen and their close circle to his cabin and ordered sailors to pour molasses for the people on deck. The Fijians thought Hudson planned to kill them all, but instead of violence, the evening passed through a series of cultural accommodations, a musical interlude to break up months of sailing and surveying. Hudson and Tui Dekreti shared cigars. When Tui Dekreti asked for his evening drink of *kava*, Hudson ordered it made from the *Piper methysticum* an expedition botanist had collected as a scientific specimen. One of Hudson's silver dish covers became a *kava* bowl. Outside the cabin, Fijians and Americans danced. Phillips led clapping Fijians in steps Americans found precise and surprisingly decent, for Polynesians. In return, as Hudson remembered it, he "endeavored to entertain them with various amusements of the men—such as singing, dancing, jumping Jim Crow." His sailors got up a "regular, old-fashioned negro entertainment." As they had often done, sailors Emanuel Howard and Simon

Fig. 21. "Vendovi," drawn by A. T. Agate, from Charles Wilkes, *Narrative of the United States Exploring Expedition* (1845).

Shepherd tapped the "dance of 'Juba.'" Ship's tailor, Ambrose Olivar, his face painted black, played Jim Crow, mounted on a jackass—two men kneeling back to back, bottoms touching. "The walking of the mimic quadruped about the deck, with its comical-looking rider, and the audience, half civilized, half savage, gave the whole scene a very remarkable effect." The Americans thought they had pleased their guests.[45]

As the sun rose the next morning, Qaraniqio's canoe appeared at the mouth of the Rewa River with Veidovi on board. (Stories report that the two brothers had passed the long night together and that Veidovi agreed to be surrendered to the Americans.) Missionaries paddled behind, come to serve as interpreters and witnesses. When Veidovi climbed on deck, sailors crowded "to stare at the miserable captive as they would at a wild beast." Expedition artist Alfred Agate and naturalist Charles Pickering recognized Veidovi as "the Feegee dandy" they had met on shore a few days earlier. Veidovi had been friendly and cooperative, posed for a portrait, and served as a "civil and attentive" guide. He had asked only for a Jew's harp, which he had given to a little girl. They confessed they had liked Veidovi.[46]

(That day, Agate sketched a first likeness of Veidovi. The pencil drawing became an engraving in Wilkes's third volume. Veidovi is poised, bearded, and beautifully coiffed. Artist Joseph Drayton, less gifted than Agate, painted a watercolor portrait about the same time. Drayton noticed Veidovi's good chest muscles and his decorated comb and gave him a stern expression. Veidovi's hair fascinated them both.)

Captain Hudson put Veidovi in "double irons," binding his hands and feet, and led him to the captain's cabin to be questioned before a "sort of council" composed of missionaries and officers. (Methodist David Cargill, a serious linguist, may have set aside his Fijian translation of Luke's gospel and taken over interpreter's tasks from Connel.) Hudson reported that Veidovi "acknowledged his guilt in causing the murder of 10 of the crew of the Charles Dogget," confessing that he had held the mate's arms while others clubbed him to death. Veidovi explained that "he had only followed the Fegee customs & done what his people had often done before!" Did Hudson ask Cargill to explain to Veidovi that new rules governed what Fijians had done before? Did the Americans actually believe he understood the charges against him and accepted the moral structures they believed belonged to Christian men?

The Americans had no legal authority to arrest Veidovi, but savage ways lacked standing as a defense in Hudson's idea of civilized practice. American contact with Fiji was intermittent, but Veidovi's capture demonstrated that American law and custom had followed the country's commercial enterprise. The law's protections extended to sailors working the sea slug trade. Hudson believed Veidovi admitted his guilt, but it is not clear that Veidovi, bound hand and foot, had much choice or that it was guilt he thought he had admitted. Hudson could have punished him, stringing him up from the *Peacock*'s cross trees. Better to be merciful, the captain decided, and take the man to America, where he would learn "all about us, how rich & great a people we were, & how, by a peaceful & honest intercourse, every thing that a Fegee man wanted would be brought to their shores, &c, &c." If he lived, Veidovi would be the most sensational of the expedition treasures. Hudson promised Veidovi's relatives he would send him back "a better man and with the Knowledge that to kill a white person was the very worst thing a Fegee man could do."[47]

Fig. 22. "Veindovi, Principal Chief of Rewa, one of the Fiji Islands. Prisoner on board U.S.S. Vincennes," drawn by Joseph Drayton, from the Logbook of the Ship *Vincennes*. (Photograph courtesy of the Peabody Essex Museum, neg. A4815.)

Reports leave a sense that Veidovi seemed to Americans more a wild animal to be tamed than a man to be taught. Still Americans thought their decision to transport Veidovi was merciful. They thought they had impressed his brothers: "The King & Phillips said that we were doing what was right in taking their Brother away, that they did not blame us, and that now they were satisfied that the thing had been accomplished in the easiest manner. They were quite reconciled to the Captain, & while they expressed much sorrow at losing Vendobi, they were no longer angry, but very sad." Hudson gave out the promised presents. Tui Dekreti held Veidovi's arms, touched his nose, and promised to take care of his fifty-five wives (an impressive number, but monogamous Americans sometimes lost count beyond three or four); Phillips, who lost an ally in Fiji's political intrigues, paced back and forth; "inferior chiefs" kissed Veidovi's hands; common people kissed his feet; a servant asked to come along but received "only a parting blessing from Vendovi, who placed both his manacled hands on his head." The day's performances ended with "the singular spectacle of a group of naked savages in Tears."[48] For Hudson and his crew, the capture of Veidovi played as good colonial theater: the defeated Fijians wept, while applauding the wisdom and kindness of the victorious Americans.

Over the next few weeks, the Americans pieced together the story of the murders and learned something of Fiji's recent political history. Veidovi's English-speaking brother Phillips described quarrels among Fiji's ruling families, drawing plot lines for the Americans that tangled through feuding families like a Shakespearean tragedy. In recent decades, Fiji had seen regicides, fratricides, plotting queens, feigned deaths, corpses made to seem alive, perfidy, treason, betrayal, hate, murder, and revenge. In 1827 Veidovi had taken a bribe from one faction and shot his brother Tui Sawau, midway through a meal at Phillips's house. "Four balls passed through his breast, but such was the strength of his constitution, that he survived for eight days." Ill will survived Tui Sawau's death, and some in Fiji supported Veidovi's capture.

In fact, while the Americans congratulated themselves on their cleverness, their justice, and their mercy, Fijians may have imagined they had used American plans to serve Fijian ends. For some, Veidovi's arrest fulfilled a prophecy that had been circulating in Fiji at least since 1837. A blind seer predicted the fate of Rewa's royal brothers: "*That one would die a natural death, another would float away, two would be killed, the most diminutive of the whole would be made king, and the principal chief of Bau would be shot during a war with Rewa.*" Veidovi had floated away.[49]

Hudson continued to press Veidovi, who had been joined by his English-speaking barber "Oahu Sam." He sat the two before another tribunal of ship's officers and asked Veidovi to confess again. This time, Veidovi explained that he and his brother Qaraniqio and a local chief at Kantavu plotted to murder the crew and seize the *Charles Doggett*. He named the Kantavu chief as the conspiracy's leader, adding that this man organized his townspeople to weave vine ropes to haul the brig on shore. Veidovi reported that the plotters had eaten a black crewman, although Veidovi "himself did not partake." No one mentioned a tobacco taste.[50] Veidovi also confessed that he had helped engineer the capture of the *Nimrod*, an Australian whale ship whose crew had been ransomed at Kantavu in 1838 for fifty whales' teeth, four axes, two plates, a case of pipes, a bundle of fishhooks, an iron pot, and a bale of cloth. The confession is a small note in a tangled record, but fifty whales' teeth would have given Veidovi enormous influence in Fiji's power plays. Those whales' teeth may have changed the political calculus among the brothers and led them to think they were better off without Veidovi.[51]

Hudson gathered more evidence a month later, when he took the testimony of James Magoun, the American working for Bachelor who survived the attack. Magoun confirmed Connel's story (although he counted ten killed, one eaten), fingering Veidovi as the one who "ordered his people to murder all the whites" (a group that included at least one man the Americans called black). Magoun swore he heard Veidovi's voice, although he did not see him speak.[52] Hudson asked Magoun to take an oath, to add weight to his insubstantial evidence. Magoun signed his statement; Hudson witnessed his signature; three officers witnessed the signatures of Magoun and Hudson; and J. H. Eagleston, the bêche-de-mer trader, tied it all down by certifying that Magoun was a "man whose statements are to be trusted." An able lawyer would have ripped the testimony to shreds, but it was enough for Hudson to justify keeping Veidovi captive. Had the winds been right, Hudson said he would have sailed to Kantavu and burned the town to punish Veidovi's co-conspirators.[53]

The winds carried him instead to a rendezvous with Wilkes, where he handed his captive over to the expedition commander. Veidovi spent the next twenty-three months, the last months of his life, on the *Vincennes*. (Pickering noticed that Veidovi measured the duration of this extraordinary rendition by counting moons.) Veidovi's transfer from the *Peacock* to the *Vincennes* passed without commotion. A sailor's note: "He is about thirty-five years of age, tall and rather slender, and has a countenance which belies his character—its expression is mild and benevolent. He was placed under the charge of a sentry, with orders

Fig. 23. "Vendovi," drawn by A. T. Agate, from Charles Wilkes, *Narrative of the United States Exploring Expedition* (1845).

not to allow any one to speak to him. At the same time the master-at-arms was called up, and directed to see that he received his meals at the proper time." Someone measured Veidovi: he was 5′11″; his foot was 11½ inches long; his arm measured 34 inches. His head had a circumference of 22 inches, and his face sloped at an angle of 67°. He had 32 teeth and a resting pulse of 65 beats per minute.[54]

Wilkes had a ship's barber trim Veidovi's hair; the barber gave locks to his friends. Veidovi was "chopfallen," Wilkes wrote, mortified at "having his huge head of hair well cropped off." This crestfallen man, wistful and clean-shaven, appears in Agate's second portrait. In this portrait Veidovi's face matches Pickering's description of a countenance he found "grave and peculiarly impressive; and I had frequent occasion to remark, that strangers did not readily forget the features of Veindovi."[55]

"Marked in Blood"

The next few weeks must have been horrible for captive Veidovi. Before the Americans left Fiji, he learned that the cultural exchanges, the drinking and dancing that had taken place on the eve of his arrest, and even Hudson's mild punishment were not to be the norm for American-Fijian meetings. Seven weeks into Veidovi's captivity, whatever peace held between Americans and Fijians broke down, and he saw for himself the face of civilized anger and the consequences of Fiji's incorporation into a new world system. Veidovi learned that in this new world the kind of everyday violence, affronts, and misunderstandings that punctuated Fiji life (Americans called it savagery) could bring on the wrath of nations.

On July 24, 1840, a group of men, assigned to one of the small crafts Americans used to conduct their surveys, went ashore to bargain for pigs and yams on the island Malolo, just east of Viti Levu. The men were actually short of supplies, but the events at Malolo illustrate how easily transactions could go wrong. The Americans had taken a young Fijian on board to hold as hostage while their comrades tried to trade for food. When the young man jumped overboard and began to run through the shallow water, the Americans fired a musket over his head. Accounts suggest that the Fijians on shore thought the Americans had killed the hostage, and they then turned on the negotiators, clubbing two young officers, Midshipman Wilkes Henry and Lieutenant Joseph Underwood. Naval men were often sentimental about dead comrades, but affection for Henry and Underwood was particularly strong. "Two officers more dearly and more justly beloved could not be found among us; and it would be difficult to find in any service, men more blameless in every point of character and conduct" summed up the feelings of many. (If the story needed an extra dose of sentiment, Henry was the only child of Wilkes's widowed sister and Underwood left widowed a woman he had married on the eve of the expedition's departure.)[56]

Americans had worn through the patience they displayed when they took Veidovi. Now Fijians would witness a different kind of American performance. When the violence at Malolo began, Veidovi was still in irons on board the *Vincennes*, but the captive chief could have had a firsthand account from his barber Oahu Sam, who was taken along to Malolo to interpret for the Americans.[57]

Wilkes collapsed in grief when he learned that his nineteen-year-old nephew was dead, but he recovered to orchestrate the American response to the "outrage": the two deaths that the Americans called "the massacre at Malolo." First, Wilkes had to dispose of the bodies, a tricky thing for Americans, who believed that Fijians made meals of freshly buried corpses. He chose a grave site, sheltered by a grove of ficus trees, on a tiny sand island, "a place far enough removed from these condor-eyed savages to permit them to be entombed in the earth, without risk of exhumation." Captain, officers, and twenty sailors, all dressed in white, formed a funeral procession, carrying the bodies to a newly dug grave. The canvas around the bodies "was crimsoned with their blood," Midshipman Reynolds remembered. Expedition artist Alfred Agate read a Methodist service, and sailors fired three musket volleys, sending island birds screeching skyward. Walking back to their ships, the Americans erased their presence, dragging palm fronds over their footprints. Wilkes regretted that he could not mark the graves on the island, but he labeled the spot on his Fiji charts—"Henry's Island," part of the "Underwood Group."[58] (The names did not stick.)

Since the Americans had burned the Fiji town of Tye on July 14 as punishment for "mere theft" (Fijians had stolen a boat and equipment worth about $1,000), Wilkes needed something harsher for Malolo. (Americans were not the only ones to burn towns; the French had done so too, although witnesses worried that town burning had little effect on Fijians, who "can build better in a week.")[59] The loss of Henry and Underwood had Americans spoiling for a fight. Sailors spent the night on deck, filling cartridges and sharpening cutlasses; on this one occasion, they appreciated their commander's inclination "to punish first & inquire afterwards." In his report on the events, Wilkes said he tried to "restrain desire for revenge within the bounds of prudent action," but he knew he could not let the people of Malolo go unpunished.

Americans landed in the morning and stormed Malolo's two villages, one left undefended, the other heavily fortified. Sailors and marines quickly overcame the defenses, firing rockets at thatched roofs and sending houses up "like tinder and the flames spread with very great rapidity." "The whole village was a mass of flames; even the cocoanut

Fig. 24. "Henry's Island," drawn by A. T. Agate, from Charles Wilkes, *Narrative of the United States Exploring Expedition* (1845).

trees scattered through had taken fire, and long streamers every now and then shot upward, kindling the thick crown of foliage around their tops and in an instant subsiding again." Bamboo cottages "exploded one after another with a sound like that of the discharge of musketry." Following Wilkes's orders, Americans destroyed "every store house for provisions . . . and as far as possible every thing that could serve as food; it being the object as much as might be, to make the island desolate," a medical officer remembered.[60]

Fijians fled the burning villages, carrying the dead and wounded on their backs. Hot air scorched the lungs, sailors complained, but they entered the villages and began sifting the ashes for the remains of dead islanders and for anything that had belonged to Henry or Underwood. They found the bodies of a man shot through the head, a woman shot "by mistake," and a baby burned to a crisp. They collected dozens of the bows and arrows, clubs, and spears that became part of the expedition's ethnographic collections, items easier to gather as spoils than to purchase from Fijians reluctant to part with them. Underwood's hatchet-slashed cap turned up. The next day a woman brought Underwood's melted watch and Henry's cracked eyeglass to Wilkes.[61]

She also carried a white cock, a peace offering Wilkes rejected "because I had no idea of making peace with them until it should be sued for after their own fashion." To Wilkes, this meant that all the survivors on Malolo must come and "beg pardon and sue for mercy; and that if they did not do so, they must expect to be exterminated." Wilkes does not tell us how his interpreter (perhaps Veidovi's barber Oahu Sam) explained extermination to the Fijians. Expedition philologist Horatio Hale offers a possibility in his "Vitian Dictionary," translating the phrase *sa yavu sara* as "it is entirely destroyed." By mid-century, extermination carried this sense of "utter destruction" among English speakers. Historians suggest that the idea of extermination would have been familiar to Wilkes's American readers, who had begun to see the word used to describe their countrymen's encounters with native peoples of the West. In Fiji the people of Malolo got a swift lesson in new usage. When American newspapers ran the story six months later, they described "A Small War." No Americans died in the fighting, but Malolo dead numbered near ninety. The violence on Malolo was one of the events that prompted eloquent Midshipman Reynolds to write, "It seems to me that our path through the Pacific is to be marked in Blood."[62]

The people of Malolo surrendered the next day, crouching on hands and knees and moaning and wailing. This was the acknowledgment that Wilkes had been waiting for. They were "our slaves, and would do whatever I desired," he said. With their heads still bent to the ground, they declared they would never again "make war against the white men." Wilkes lectured the prostrate crowd, explaining the atrocity of their crime and the justice of his actions. "I told them they might consider themselves fortunate that we did not exterminate them." He ordered the survivors to bring food and water to the Americans, dismissing Fijian complaints that sailors had burned all the island's yams and coconuts. The people carried 3,000 gallons of water to the American vessels and surrendered 12 pigs, 3,000 coconuts, and 3,000 yams. (Since they surrendered these yams at the end of July, just about the time they would be planting a new crop, the survivors of Malolo must have had hungry months waiting for the next spring's yams to ripen.)

From the distance of more than a century and a half, the American response to the murders seems out of proportion, even as an expression of the enormous grief felt by sailors, vulnerable and far from home. Wilkes had to answer for his actions when disgruntled officers included the use of force at Malolo in the charges of misconduct they brought against him back in the States. The court dismissed complaints that the destruction at Malolo exceeded the official orders, which permitted the expedition to employ force "merely for its necessary defense." Those orders described expedition purposes as "altogether scientific and useful, intended for the benefit equally of the United States and of all commercial nations of the world." "Its path upon the ocean will be peaceful, and its pursuits respected by all belligerents." In his defense, Wilkes insisted that he urged moderation on his angry sailors, even though American deaths in Fiji called "for retribution, and the honour of our flag demanded that outrage upon it should not remain unpunished."[63]

Although it risks turning a sad story into an abstraction, there is evidence to suggest that Wilkes and his men turned quickly to violence because people in Malolo finally behaved as Americans expected Fijians would. The kindness, hospitality, and cooperation that characterized the first two months in Fiji confused the Americans, just as the mild expression on Veidovi's face confused sailors who wanted to see a savage. Pickering confessed in his diary on July 3, 1840, "We had hitherto been so well treated by the Natives, had found them always so obliging, & so 'timid'; that many of us began to think they had been maligned. Some even doubted whether they were really *Cannibals*; & the matter [the word "question" is penciled above] had been seriously discussed at the Wardroom table the previous evening. It so happened that though we had been nearly two months at these islands, no one could say that he had actually witnessed the fact, or name a person of credit who had. We were on the point of returning, & only adding 'mud to the stream of knowledge.'" When some men on Malolo finally behaved like Fijians and killed the two American officers, it cleared the mud and let loose a stream of violence. One group on Malolo bore the brunt for past Fijian wrongs, real and imagined.

Wilkes spent the remaining weeks on Fiji in a naming frenzy, labeling small Fijian islands for his officers. He penciled in the Hudson Isles and the Ringgold Isles for the captain of the *Peacock* and the captain of the brig *Porpoise*. He named a river for artist/naturalist Peale, and peaks for "scientifics" Hale and Pickering and for artist Drayton. Agate got an island. These Americans were happy to sail off, leaving only their names behind. "On taking our final departure from these islands all of us felt great pleasure; Vendovi alone manifested his feeling by shedding tears at the last view of his native land."[64]

The Fejee Mermaid

When the peaks of Fiji sank below the horizon, Wilkes removed Veidovi's irons and lifted the ban on conversation. Sailors recorded Veidovi's comments. He asked about the surveyors. They made "pictures of the Land, so that other ships would know where to go as well as the natives, who lived on the Islands," Reynolds remembered telling Veidovi, an explanation certain to trouble a man worried about a future when Americans and Europeans moved as freely as Fijians through Fiji's islands.[65] The white settlement at Oahu impressed him, so did the order and discipline on American ships, but no one explains how the tall Fijian fit into the work rhythms of the ship. He sometimes talked to the scientifics, telling Pickering that young lovers in Fiji ran off when a man was too poor to pay a bride price and describing a great flood that had destroyed the islands to philologist Hale. A Pacific tsunami, Hale thought. Veidovi confirmed American hunches about the superiority of white people, expressing "contempt for the colored race" and complaining about the filthy people he saw in Oregon when, dressed in "Fiji fashion," he left the *Vincennes* to parade with American sailors celebrating Independence Day 1841 along the Columbia River.[66]

When the *Vincennes* docked in San Francisco in August 1841, Californians came on board to shake hands with the captive; "he returned the salute with a kind of nod, showing some appreciation of the attention paid him," a man remembered. Veidovi appeared "impatient and uneasy," and even though he had lost nearly fifty pounds in captivity, at two hundred pounds he still cut an impressive figure. When the homeward-bound expedition stopped in Honolulu in November, children climbed aboard the *Vincennes* to see the "cannibal chief," a "brawny giant" to them, but one "whose sullen countenance clearly revealed the misery of his punishment."[67]

Misery is about as close as we can get to a word for Veidovi's last months. Wilkes, who was often lonely in his command, sympathized with his captive, but there was little he could do to ameliorate the man's condition. Years later Wilkes remembered that Veidovi "conducted himself with much propriety and was inclined to assimilate himself with the officers to whom he believed he was entitled to associate with." He was "a model of a man, very tall and erect and of a proud bearing, scrupulously clean in his habits." Wilkes honored his captive, naming an island in Puget Sound "Vendovi Island."[68]

Chroniclers agree that Veidovi's health declined quickly in early spring 1842, particularly after Master's Mate Benjamin Vanderford died as the *Vincennes* passed beyond Java and headed west across the Indian Ocean. Wilkes remembered Vanderford as a good officer. The Salem-born bêche-de-mer trader and ship's pilot, a man in his mid-fifties, had come to Fiji with the first Americans around 1809. He knew Fiji well, spoke the language, and had Fijian friends who called him "Put-e-mum." In the long months at sea, he and Veidovi became friends, and some witnesses say that Vanderford promised to stand by Veidovi in America.

Details of Vanderford's death leave little wonder that it sent Veidovi sorrowing. Charge of the ship's "spirit room" where liquor was stored proved an unhappy assignment for Vanderford, and Veidovi watched his friend drink himself to death. On March 20 Vanderford suffered his first attack of "horrors" brought on by the alcohol-induced delirium that would do him in. He raved at shipmates, accusing them of "pricking him with sharp wire and mocking him with strange gestures." He tried to wave away the men in red nightcaps he saw crowding into his room. He shouted that Captain Wilkes was coming to flog him, and then rushed around his cabin, madly packing his things into a bag. Shipmates tried to calm him with a dose of laudanum. He died in his sleep. A quick postmortem revealed "an extensive disease of the brain," although the surgeon could not be certain the brain disease was actually extensive enough to have killed him. Shipmates read the Methodist service and slid Vanderford's body into the Indian Ocean. They fired three musket volleys over the sinking corpse. Veidovi refused to look at the body.[69]

The real work of the expedition was over, and the *Vincennes* headed back to the States. The ship anchored off Cape Town in April, stopped in St. Helena in May, where a sailor picked flowers from Napoleon's grave and pressed them into his journal, and arrived at New York's Sandy Hook at noon on June 10, 1842. Wilkes disembarked from the *Vincennes* and sent the dying Veidovi to the Naval Hospital, where expedition surgeon James Croxall Palmer nursed the Fijian through his last hours.[70]

If Wilkes hoped to be hailed as a hero on his return to the States, he was disappointed. Even though he escaped his court-martial with only a reprimand for illegally flogging sailors and marines, his prospects for triumph looked dim in the short term. Little in contemporary American politics or culture would have bolstered his confidence in his expedition's place in history. When the ships sailed from Virginia in 1838, the U.S. Ex. Ex. carried the hopes and scientific commitments of the Democratic administrations of Andrew Jackson and Martin Van Buren. By the time Wilkes returned to New York in 1842, Whig John Tyler had been president for a little more than a year. (Tyler took office on April 4, 1841, when William Henry Harrison died after a quick bout with pneumonia.) A slaveholding aristocratic Virginian, Tyler was hardly a model Whig, but whatever the president's politics, Wilkes found that Tyler had little interest in celebrating the triumphs of the previous administration's scientific exploration. Historians also speculate that celebrating Wilkes's accomplishments in the Pacific, particularly his survey of the Oregon coast, risked spoiling delicate negotiations with England over the boundaries of the Oregon Territory.[71]

Technology was changing too. Wilkes led one of the last circumnavigations under sail. In fact, steam launches brought his men in and out of major ports. He sailed with sketch artists, but he was astute enough to note that when he returned, daguerreotypists, capitalizing on the new French invention, had opened parlors in New York. It would take time for expedition artists like those in Wilkes's employ to learn the new technology, but the writing was on the wall. Sadder for Wilkes were the events in the country's march across North America in the 1840s, which overshadowed the expedition's accomplishments. For land-hungry Americans, reports of the discovery of an ice-covered continent and surveys of scattered Pacific islands paled beside news of a treaty that guaranteed to the United States control of a large portion of the Oregon territory and victory in a war with Mexico that doubled the size of the country. New lands in Oregon and California were as rich and diverse as the lands Wilkes and his crews had surveyed and far more promising for the near future of American settlement. His Pacific discoveries were shelved while the American empire worked its way west. But his work was not forgotten. In the 1940s, the U.S. Navy sailed against the Japanese with Wilkes's century-old charts.

Even the stories written and published by Wilkes and his men faced competition from the tales of the new American West. Wilkes had instructed each of his officers to keep a "journal during the cruise, which they will send to the commander of the ship to which he may be attached, weekly." He offered a practical explanation for the order, suggesting that it was important for each officer to know the actual situation of a ship at any given time, but he was also interested in "the habits, manners, customs, &c., of the natives and the positions, descriptions, and characters of such places as we may visit." As a result of Wilkes's order, there is an extensive record of the expedition, although officers' written opinions may well have been tempered by the promise of weekly scrutiny. Nevertheless, when Wilkes wrote up his *Narrative of the United States Exploring Expedition* in the mid-1840s, he had a number of accounts to draw on, and he worked reports from several witnesses into his five volumes. Though rich in detail, Wilkes's narrative lacks the drama and artistry readers found in accounts of continental adventurers like John C. Frémont, agent and chronicler of America's continental expansion. Frémont's stories eclipsed material from the U.S. Ex. Ex., in part because Frémont had political connections and literary gifts (thanks to his wife, Jesse Benton Frémont, daughter of a powerful senator and a very good writer) that simply surpassed any Wilkes could command.[72]

Even though literary fame eluded him, Wilkes had reason to hope that his scientific and ethnographic collections would secure his expedition's place in history. President Tyler's welcome was chilly, but Wilkes discovered allies in Washington among the men who had formed the National Institution for the Promotion of Science. Wilkes learned of plans for this National Institution when the squadron stopped in Honolulu in the summer of 1840 and immediately sent a letter asking to join. In the fall of 1842, fellow members crowded a lecture hall to hear Wilkes speak. For most Americans, however, a more eloquent record of the expedition was stored in the things Wilkes and his men collected—the plants and animals, weapons, cloth, and baskets—those marvels Meigs had seen on his way to visit Veidovi's rare and unusual specimen skull.[73]

There is no question that had Veidovi lived, he would have caused a stir, particularly in New York. In the 1840s, Native American performers still drew crowds in Europe and America, but few in those crowds (with the possible exception of whalers back "from the Fejees") could ever have seen a man like Veidovi. Veidovi died after just a few hours in New York. But his story left its mark on American culture. Through the month of June 1842, *New York Herald* publisher James Gordon Bennett dangled the dead "cannibal" before his readers, milking what he could from the story through the last weeks of June. Corresponding papers reprinted jokes that Veidovi was "sick with consumption in consequence probably of having no human flesh to eat." The *Southern Literary Messenger* reported that "poor Vendovi was finally attacked with a disease (common, we believe, among wild animals in a state of confinement), ossification of the lungs, and whilst being removed from the ship to the hospital, raised his head and exclaiming, 'I feel that the great spirit has sent for me,' uttered feebly his war cry and sunk back dead in the bottom of the boat."[74]

Master showman P. T. Barnum, proprietor of the American Museum, experimented with how best to capitalize on public fascination with Veidovi. On June 27, 1842, Barnum inserted an announcement in the *Herald*: "By permission of the U.S. Government, the head of Vendovi, the Cannibal Chief, from the Fejee Islands, has been deposited in the museum for public exhibition, for one week only." It is understandable that those who rushed out to see the head without reading Barnum's entire paragraph were disappointed. A last sentence qualified his claims: "It is a true & exact cast of this sailor-eating chief taken by order of the government, by Thomas Holmes, Esq. the celebrated artist of this city." (Disappointed patrons could console themselves with Barnum's other attractions: "Industrious Fleas"; Mr. Boyce, the comic melodist; Mr. Brooks, the unrivaled dancer; Miss Rosalie, the charming vocalist; La Petite Celeste, the beautiful danseuse; a glass blower; an albino lady; a fortune-telling gypsy girl; and 500,000 curiosities.)[75]

Ever resourceful, Barnum didn't need an actual head. Curiosity about Veidovi triggered his imagination. On June 18, 1842, with the press still reporting on the "dead Fiji cannibal," Barnum and a partner bought an old stuffed mermaid from a Boston sea captain. The Boston man inherited the mermaid from his father, he said, who purchased it from a man in Batavia sometime around 1820. His father believed that a Japanese fisherman, trolling in the Sea of Japan, had found a mermaid squirming among the fish in his nets. Confident that a dead mermaid would make his fortune in London, the sea captain mortgaged his ship to buy the creature. Man and his mermaid arrived in London in 1822, and, as he had hoped, his strange little creature, hideous and contorted under a glass globe, became the talk of town. Naturalists and anatomists weighed in, exposing the mermaid as the work of a clever taxidermist who had stitched the tail of a salmon onto the head and torso of a baboon (or an orangutan). But expert opinion only fueled public interest and more people paid to see the mermaid. Science or hokum?

The sea captain and his creature traveled a circuit of English fairs for a few years but dropped from sight about 1825. Historians believe that this London mermaid, a bit shabby from its travels, turned up in Boston in 1842. Like previous owners, Barnum stoked public curiosity by stirring up controversy among "learned naturalists" he invited to speculate on the mermaid's authenticity. But it was Barnum's genius to connect the little withered creature to the man from Fiji who had just died in New York. In late June, Barnum began to advertise the mermaid as a "great curiosity," "taken among the Fejee Islands" and preserved by Chinese arts. The creature's associations with Japan and Indonesia disappeared, and for the next few years Barnum's "Fejee Mermaid" was an American sensation. Barnum and his partners kept up the pseudo-scientific chatter about the creature, and eight years after Veidovi had died, Americans were still buying tickets to see "The Remarkable Fejee Mermaid." The mermaid helped the showman earn his spot among the nineteenth century's immortals.

No surprise that phrenologists, always ready to take up a notorious head, seized on Veidovi's story too. Writing in 1847, popular phrenologist Orson Squire Fowler found "cautiousness" small in "Vendovi, the Feejee chief." Youths be warned, he wrote, that such a man "fears nothing" but is "perpetually in hot water." Fowler's lessons and Barnum's creature honor this man who missed his chance for a proper Fiji burial.[76]

Empty Graves

This complicated encounter between Americans and Fijians ends with trouble over bodies of three men—two Americans and a Fijian—each of whom died far from home. It is certain that the deaths on Malolo of Wilkes Henry and Joseph Underwood hit the men of the expedition hard. Comrades buried their bodies and avenged their deaths, but they acknowledged that their work was incomplete until they found a way to accommodate the needs of stateside mourners—in this case, mothers, widows, and friends. Graves were particularly important to this generation of middle-class Americans, who perfected the arts of mourning and burial and changed their cities' landscapes by building lovely garden cemeteries, like Laurel Hill in Philadelphia, where Morton was buried.

The officers of the expedition shared the cultural assumptions of

Wife Murdered by her Husband.

Also, groups of a School,—a Milliner's Shop,—a Shoemaker's Shop,—a Barber's Shop,—a Blacksmith's Shop, &c., as well as a variety of single figures of distinguished men, &c., with upwards of

One Hundred of Cabinet Size,

rendering it the largest collection of Wax Figures in America.
The entire of the above, and the immense collection of

BIRDS, BEASTS, FISH, INSECTS & REPTILES,

obtained from all parts of the world, together with innumerable varieties of Natural and Artificial Curiosities;

Paintings, Engravings and Statuary,

OIL PORTRAITS

of the GREAT and GOOD of all nations,—Naval and Military Heroes, Patriots, Statesmen, and Divines;—Rare Coins and Medals;—Shells, Corals, and Fossils;

EGYPTIAN MUMMIES,

and ancient Sarcophagi, 3000 Years old, and an entire

Family of Peruvian Mummies:

the DUCK-BILLED PLATYPUS, the connecting link between the BIRD and BEAST, being evidently half each;—the curious half-fish, half-human

FEJEE MERMAID,

which was exhibited in most of the principal cities of America in the years 1840, '41, and '42, to the wonder and astonishment of thousands of naturalists and other scientific persons, whose previous doubts of the existence of such an astonishing creation were entirely removed;

Elephants and Ourang-Outangs;

ANIMALS and BIRDS of every nation; Sharks, Seals, and a variety of FISHES, including the curious

SAW AND SWORD FISH,

all in lifelike preservation; the whole forming a School of Instruction, blended with Amusement, that for extent and interest is unequalled in the known world;—the whole to be seen for the small admission fee of

TWENTY-FIVE CENTS.

In Addition to which, and

WITHOUT EXTRA CHARGE,

visitors are admitted to the gorgeous Exhibition Hall, which has been newly decorated at an expense of nearly five thousand dollars, where they can witness the magnificent

THEATRICAL ENTERTAINMENTS,

given EVERY EVENING, and WEDNESDAY and SATURDAY AFTERNOONS, by a Company of Comedians and an Orchestra of Musicians, admitted to be SUPERIOR to any ever before collected in this country, with the aid of

Stage and Scenic Arrangements,

the most grand and superb ever seen in either Europe or America! thus warranting the universal admission that the Boston Museum, besides being the most comfortable and genteel, is also the

Cheapest Place of Amusement

IN THE WORLD! A single visit will prove the truth of this assertion, as the admission is only

☞ 25 Cents to the Whole!!!

Fig. 25. "Fejee Mermaid," from the *Norfolk (NH) Democrat*, October 28, 1853. (Courtesy of the American Antiquarian Society.)

their more sedentary relatives, and they struggled to find the right way to memorialize their lost companions. Wilkes had penciled the names of Henry and Underwood onto his charts, but no grieving family member was likely to venture to a grave site hidden in a grove of ficus trees on a small sand island somewhere in Fiji. The officers had to do something. On August 8, 1840, shortly before the expedition sailed from Fiji, officers on the *Vincennes* passed five resolutions honoring Henry and Underwood and pledged $2,000 out of their salaries to erect a monument over an empty grave at Mt. Auburn Cemetery in Cambridge, Massachusetts, the first of America's garden cemeteries. Artist Drayton designed a cenotaph to honor the two who had fallen "by the hands of savages, while promoting the cause of science and philanthropy." He also added the names of two men whose lives were lost in icy waters when the expedition rounded Cape Horn.[77]

There is no way to know how Veidovi imagined his fate as he watched the islands of Fiji disappear. Or how his family and friends pictured what was to happen to him in the United States. Perhaps a sailor brought the news that Veidovi had died in New York, and the blind prophet took it as a sign that his prediction that one brother would "float away" had come true.

But what of Veidovi's corpse? It is possible that the journal-keeping American officers who had described the elaborate funerals for Fiji chiefs acknowledged that it was unlikely that their captive would be able to return home to die and be buried. Those who had written on Fiji burial customs knew that as Veidovi lay dying in Brooklyn, he had no friends to give him whales' teeth to throw at the tree that blocked the path to the regions of the dead. No friends to lament over his corpse. No priest to select two grave diggers and give them mangrove staves to dig a grave beside a stream. No grave diggers to wash his body, anoint it, dress and decorate it as if "for a great assembly of chiefs." No female friends to kiss the corpse; no wives, dressed in their best clothes, to come forward to be strangled and buried with him. There would be no great feast at the end of ten days when the taboos were lifted from those who had assisted in the burial. No one to mark the moment when the soul of the deceased could "quit the body and go to its destination."[78]

Instead, Veidovi's skull went to Washington, where its presence "has been a matter of public record since at least 1846," as a physical anthropologist wrote in the late 1960s. There is some evidence that the crania from the Wilkes expedition were stored temporarily in Philadelphia in

(Front.)
TO
THE MEMORY OF
LIEUTENANT JOSEPH A. UNDERWOOD,
AND
MIDSHIPMAN WILKES HENRY,
UNITED STATES NAVY.

(Rear.)
LIEUTENANT UNDERWOOD,
AND
MIDSHIPMAN HENRY,
FELL BY THE HANDS OF SAVAGES,
WHILE PROMOTING
THE CAUSE OF SCIENCE AND PHILANTHROPY,
AT MALOLO,
ONE OF THE FEEJEE GROUP OF ISLANDS,
JULY 24, 1840.
PASSED MIDSHIPMEN REID AND BACON
WERE LOST AT SEA, OFF CAPE HORN,
MAY, 1839.

(Right.)
THIS
CENOTAPH
IS ERECTED BY THEIR
ASSOCIATES
THE
OFFICERS AND SCIENTIFIC CORPS,
OF THE
UNITED STATES
EXPLORING EXPEDITION.

(Left.)
TO
THE MEMORY
OF
PASSED MIDSHIPMEN
JAMES W. E. REID,
AND
FREDERICK A. BACON,
UNITED STATES NAVY.

Fig. 26. "To the Memory of Lieutenant Joseph A. Underwood and Midshipman Wilkes Henry, United States Navy," designed by Joseph Drayton, from Charles Wilkes, *Narrative of the United States Exploring Expedition* (1845).

the spring of 1851, just before Morton died. We know for sure that in the following years, Veidovi's skull moved around Washington museums, following intellectual developments in American ethnography and physical anthropology. As Meigs learned, the skull was first displayed among the expedition's ethnographic materials at the National Institution for the Promotion of Science. In 1856 the Smithsonian Institution absorbed those collections, but Veidovi's skull was a strange fit in the Division of Mammals, where it was first housed.[79] In 1868 Veidovi's skull was transferred to the Army Medical Museum, an exotic object in an anatomical collection bulging with body parts culled from America's own battlefields. Curators at the Army Medical Museum measured the skull and labeled it specimen 292, dropping its sensational ties to cannibalism and murder. In their catalogs, it was "(**292**) Cranium M. æt. C. 40, Cap. 1495 c.c., L. 187 mm., B. 141 mm., H. 136 mm., I.f. m. 49, L. a. 391 mm., C. 467 mm., Z.d. 145 mm., F. a. 70°. 'Vendovi,' chief of one of the Fiji Islands." Curators cleaned off the skull's last traces of flesh and hair, leaving a residue of an alkaline macerating agent coating its bones, and then inked the numbers "292" onto the forehead, giving Veidovi a permanent place on the "List of Specimens."[80]

Veidovi's skull moved again in 1898, when the Smithsonian opened its Division of Physical Anthropology. For physical anthropologists, Veidovi's skull was not a cannibal sensation or an exotic artifact from the Wilkes expedition, but a record of Fiji's racial and ethnic past. Veidovi may have had Tongan, Samoan, and Fijian ancestors, but physical anthropologists appreciated his skull because it came from a man born before the numerous mixed-race offspring of fertile characters like Paddy Connel had a chance to spread their European genes through Fiji, before the English imported workers from India to man plantations of the British "protectorate" in the 1870s.

Detailed description of one skull was not much to go on. Measurements place it between Melanesian and Polynesian populations, confirming what a nineteenth-century observer might have concluded from studying Veidovi's portraits: he had what contemporaries called Melanesian hair and a Polynesian face. No statistical case could be made from Veidovi's skull, but his skull's measurements carry traces of stories of migration, of love and sex, that the man never had a chance to tell. We do know that cultural changes, impure encounters between people from different parts of the world, carried this skull to New York, then to Washington, where, strangely, it came to represent biological purity.

It is hard not to marvel at Veidovi's unlikely trip around the world to America and at the history of his skull. We have some of the facts of his story and three images of a face that strangers did not readily forget. We have his physical dimensions and one skull. This one skull cannot close the gap between the United States and Fiji, yet the skull's preservation in an American collection has created possibilities for a part of Veidovi's history to return to Fiji. Political shifts in Fiji delivered him once into American hands; twenty-first-century shifts in America and in Fiji may return his skull to Fiji. But we do not yet know how this skull with its distinct past will play into the unfolding history of Fiji's present racial politics, although descendants have come to visit and accepted their ancestor's place in the American archive. (In the early 1990s, the American State Department got involved in the skull display and asked the Smithsonian to return Veidovi's remains to Fiji as a "gesture of respect and friendship," even though no formal request had come from Fiji.)[81]

The twists in this story suggest some of the difficulties controlling the ideas unleashed by the unburied dead. Although Veidovi's skull sits peacefully in Washington, in Fiji the skull might yet become a prop in movements that claim special status for indigenous Fijians. It would be ironic if Fijians adopted nineteenth-century America's racial categories to assert their rights. Should Veidovi's skull represent a racially pure Fiji meant to appeal to fundamental nationalists who would limit political power to those able to claim status as indigenous Fijians?[82]

Or should his skull with its ties to the sea slug trade and with its traces of the exercise of Western power over Fiji's islands serve to tell a different story, one shared by indigenous Fijians and the Indo-Fijians whose ancestors came to labor on British plantations? Like that first generation of laborers, Veidovi was caught in currents that changed the world. Like many laborers born in India and working in Fiji, he died far from home. These shared histories of disruption and destruction, creativity and persistence, may yet prove more useful to twenty-first-century Fijians than the mistaken biological certainties of Meigs, Morton, and their craniologist circle.

490. Portion of bone of left leg, showing shot fracture. Primary amputation in middle third. Bullet attached. No history. Contributed by Assistant Surgeon G.M. McGill, U.S. Army.

—JOHN SHAW BILLINGS, *Description of Selected Specimens from the Army Medical Museum* (1892–93)

(**481.**) Cranium. F. æt. c. 50. Cap. 1480 c.c., L. 183 mm., B. 148 mm., H. 129 mm., I.f. m. 45, L.a. 360 mm., C. 521 mm., Z.d. 140 mm., F.a. a. 76°. Presented by Acting Assistant Surgeon G.P. Hachenberg.

—GEORGE A. OTIS, *List of the Specimens in the Anatomical Section of the United States Army Medical Museum* (1876)

(**966.**) Cranium. M. æt. c. 45. Cap. 1350 c.c., L. 180 mm., B. 134 mm., H. 132 mm., I.f. m. 39, L.a. 360 mm., C. 506 mm., Z.d. 140 mm., F.a. 76°. From Dry Creek, Utah. Presented by Acting Assistant Surgeon H.C. Yarrow. ("Wah-ker," a celebrated chief.)

—GEORGE A. OTIS, *List of the Specimens in the Anatomical Section of the United States Army Medical Museum* (1876)

CHAPTER 5

The Unburied Dead

"Precious Dust"

When U.S. Army surgeon George McGill died of cholera in July 1867, his companions buried his body beside the trail some miles east of Fort Lyon, the army's windblown outpost in eastern Colorado. Back in New Jersey, Professor Alexander T. McGill struggled through that summer haunted by his son's "precious dust" buried "without a coffin on the plains of Colorado." In November he finally wrote to Dr. John Shaw Billings at the Surgeon General's Office in Washington, D.C., to ask Billings, please, to send an army detail to dig up George's body and ship it to Princeton. He thought Billings would understand, not just because requests to move bodies often came across his desk, but also because Billings and George had been friends, comrades in the Union army's Medical Corps. In fact, the two men were so close that George had written first to Billings ("as I love and trust you most") when he wanted to share happy news that he was to be married in December 1866. Professor McGill counted on that wartime trust when he made his sad postwar appeal.[1]

Personal ties make McGill's request unusual, but this grieving father joined tens of thousands of Americans who lost loved ones during the 1860s and then worried about their "precious dust" disappearing far from home. McGill and Billings both knew that the Civil War's body count had staggered America, destroying a world where families helped relatives die at home and buried their remains close by in rural churchyards or garden cemeteries. The North and the South—though divided over slavery, united in grief for young men, and former enemies—shared the sense that their losses had been a noble sacrifice. Like his contemporaries, McGill lived troubled by images of an adult child's unburied or poorly buried body; he saw a world left out of joint not just by the too early deaths of young men but also by their poor burials. War turned grieving families and communities in the North and the South into funerary innovators, as carnage schooled a nation in the emotions of death and dying and offered practical lessons in earthly things like the preservation, transportation, and disposition of thousands and thousands of dead bodies.

George McGill and John Shaw Billings had learned lessons about bodies during their war years. The two were in their late twenties when the Civil War ended; young doctors but with medical experiences beyond their years. As a reporter for the *New York Times* had speculated in 1862, the war indeed had proved "a colossal experience" for doctors, who had dressed more wounds and seen more diseases and deaths than in a dozen lifetimes of civilian practice. In 1865 McGill and Billings staked their futures with the postwar army, but the months following the Confederate surrender pulled their lives in different directions. McGill was ordered to Fort Lyon in the Colorado Territory, and in July 1867 he and his young wife, Helen, headed west, but not into the adventurous new life they expected. They met the cholera epidemic sweeping the plains. Helen died on July 16, near Fort Harker, Kansas. George died a week later, some miles east of Fort Lyon.

That summer Billings was in Washington, at the start of a remarkable career in public health and medicine. Billings worked as an engineer of information, helping to organize the specimens in the Army Medical Museum, to establish the Surgeon General's Medical Library, to design the Johns Hopkins University Hospital, and to direct the New York Public Library. We don't usually elect administrators into our pantheon of heroes, but perhaps we should. By all reports, Billings had an exceptional gift for organizing things, whether at battlefield hospitals, in museum collections, or in libraries, for creating "order out of chaos," as he described his work on the Gettysburg battlefield.[2]

Professor McGill explained that Billings could ease the family's pain if he could just bring George's "remains away from Salt Bottom; where the freshets of the Arkansas will overflow it + wantonness of the savage may obliterate every mark that the stream will have spared. And then, who knows how soon the sacred spot, if not liable to overflow, may be a plowed field. The rail must of course end the traveling of caravan trains, very soon: + in a few months more, no one can tell where this heroic boy is sleeping." "Oh no, my dear friend," he wrote, "I cannot rest, till all that remains of George is brought, back to sleep with me, till the resurrection."[3] It is not surprising that McGill, a professor of pastoral theology at the Princeton Theological Seminary, held to a sacred faith in future resurrection. But George's death cut his connection to his country's secular creed that the plowman, the emigrant, and the railroad man carried America's future west across the continent.

McGill, like many of his countrymen, believed that a proper burial could help mend this shattered faith. He remembered that a few lucky families (if luck is the right word) had found undertakers to embalm bodies, pack them into metallic coffins, and ship them home. (That group included Samuel George Morton's widow, who buried a son—army engineer James St. Clair Morton, killed in June 1864 in the assault on Petersburg, Virginia—next to his father in Philadelphia's Laurel Hill Cemetery.) Professor McGill had seen Princeton friends and neighbors walk in sad procession to railroad stations and freight depots to retrieve the embalmed corpses of loved ones.

Professor McGill also knew that while George traveled west in the summer of 1867, army comrades in the East were busily sorting and reburying the war's dead. Thousands would never be identified. But the unnamed bodies of many who had died in camps, in hospitals, or on battlefields and been tossed into hasty graves were now given proper burials. Whispers that battlefield corpses had rotted in the open air, or been eaten by wolves, vultures, or wild hogs, or even scavenged by anatomists, fueled the country's political will. By the time Professor McGill grieved for George, a consensus had emerged that the federal government should take charge of Union war dead, and burial details, funded by a vote of Congress, were at work on the battlefields in Virginia and Tennessee, sorting the dead and arranging for reburial in national cemeteries. Northerners and Southerners shared concerns about unburied or misburied dead, although programs launched by the federal government in 1866 and 1867 focused efforts on the Union dead. Historians have argued that burial details intentionally left Confederate corpses to rot where they lay. Care for the Confederate dead emerged in a series of private efforts as women and their communities assumed responsibility for bodies and for graves. Private practices helped keep Southern sectional identity alive through that century and into the next.[4]

Mass dying and the postwar efforts to deal with the dead gave Professor McGill confidence that someone in the government could ship his boy's body home from Colorado. But the poor man came more than a year late to the national mourning, and the loneliness of his grief is poignant. The country's future had no place for George and, worse, no room for a death like his in the narratives of suffering and redemption that had begun to shape the nation's memories of the war. Professor McGill sensed that a nation exhausted by its war dead was not ready to pause for one man dead and buried in Colorado, and he regretted George's decision to stay with the army.

If pressed, Professor McGill could have explained that George's own war stories were behind his confidence in new technologies for preserving and transporting bodies—a scientific side of his sentimental letter. George may have told his father about a circular he and fellow medical officers received in May 1862, requesting them "diligently to collect and to forward to the office of the Surgeon General, all specimens of morbid anatomy, surgical and medical, which may be regarded as valuable; together with projectiles and foreign bodies removed, and such other matters as may prove of interest in the study of military medicine or surgery." Contributions formed the core of a collection for an army medical museum. "Each specimen," he added, "will have appended the name of the Medical Officer by whom it was prepared."[5]

Collectors with a philosophical bent described the museum as "one of the large compensations of human history, since it offered a means to extract beneficial knowledge from one of the periods of pestilence and war with which our race is scourged."[6] Museum planners had a British model in mind. The English army's medical museum documented the work of military surgeons to help educate a next generation of army doctors. These good goals did not reach everyone. Collection curator John Brinton worried that his least-educated colleagues didn't understand the purpose of the museum. And many field surgeons simply did not have time to collect body parts. "It was hard enough to be worked day and night in those great surgical emergencies, accompanying fierce and protracted battles, and it really seemed unjust to expect the rough preparations necessary to preserve for the Museum, the mutilated limbs. These were usually buried in heaps." If the heaps weren't coming to him, Brinton would go to the heaps. "Many and many a putrid heap have I had dug out of trenches where they had been buried, in the supposition of an everlasting rest, and ghoul-like work have I done, amid surrounding gatherings of wondering surgeons, and scarcely less wondering doctors. But all saw that I was in earnest and my example was infectious."[7]

It is hard to imagine that collecting pieces of shattered bodies could have been as cheery as Brinton recalled, but his earnest scavenging nevertheless inspired medical colleagues to begin collecting for the museum. With a clever touch, Brinton silenced his loudest critic colleagues by thanking a few of them publicly for specimens Brinton had collected and prepared himself. Apparently the flattery of false credit did the trick, and the museum soon had a large corps of willing donors—and sadly, no shortage of body parts as this war sharpened its acute capacity to maim, grinding through its armies with grim efficiency in the spring and summer of 1863.

Assistant Surgeon George McGill was one of the Army Medical Museum's early donors. He contributed bones from three gunshot-fractured arms he amputated in late 1862 or early 1863. A few months later, he sent along the shaft of humerus, a left femur, a right forearm, a right foot, and the bones of a left leg, "showing shot fracture. Primary amputation in the middle third. Bullet attached." If he followed museum instructions, he penciled "G. McGill" on a block of wood or scratched the letters onto a sheet of lead and attached it to each of the specimens he packed into a keg of whiskey and sent to Washington at the government's expense. His name would still be legible when the Surgeon General's staff unpacked his alcohol-pickled donations.[8] Late in the war, McGill tried his hand at preserving soft tissue and forwarded two kidneys (one "with a certain amount of fatty degeneration") he had removed from the body of "Corporal W.F., 'B' 1st Colored Troops, 20; gunshot wound left thigh, probably spring 1865; died of exhaustion after erysipelas, Baltimore, 20th February 1865."[9]

The catalog note pairs the two young men in a sad tableau, their fates linked by a brief wartime encounter: Corporal W.F., a black man, dies, his body beaten by infection; Assistant Surgeon McGill gives the corporal's kidneys to the museum and then months later leaves for Colorado. Billings could not grant Professor McGill's wish for George's body, but the museum catalog gives George his shot at an afterlife in American culture; it takes some archival sleuthing to find the tiny memorial, but there it is, a tribute to Professor McGill's heroic boy preserved in the leg bones, arm bones, and kidneys he donated to the museum.

Professor McGill was not the only man in postwar America worrying a government official about a dead body. Most people, grief-stricken, fussed about the war dead. Who would bury them? How? Where? Like McGill, some wondered who would save the graves from flooding streams and farmers' plows. But postwar traffic in bodies also dredged up older issues about race and Native America and set the stage for the moral, political, and ethical questions that have fueled the repatriation debates of the last two decades. In January 1865 the Harvard naturalist Louis Agassiz wrote to Secretary of War Edwin Stanton, asking if he could ship a couple of bodies up to Cambridge. "Now that the temperature is low enough," Agassiz wrote, "permit me to recall to your memory your promise to let me have the bodies of some Indians; if any should die at this time." If a January thaw threatened, Agassiz recommended that a surgeon "inject through the carotids a solution of arsenate of soda. I should like one or two handsome fellows entire and the heads of two or three more."[10]

Outside the university lab, curiosity sometimes took an irreverent turn. College students who spent a summer in Montana collecting Sioux skulls for the Smithsonian felt fine asking the Surgeon General's Office for "a spare human cranium" for their "college society at Yale." Color, sex, and race didn't matter; they just wanted a skull. A souvenir, a curiosity, a specimen, but not the head of an individual who had died.[11]

McGill did not collect human skulls, but his story introduces the postwar changes to work with collections of crania. Sad-eyed Morton looked down from a portrait on the museum library wall to survey a collection of skulls far larger than he could have imagined. Thousands of dead bodies and body parts moved around postwar America. President Lincoln's body, on its sorrowful journey from Washington to Springfield, is the most famous. But ordinary dead folks moved too, although their fates reveal a startling contrast in postwar American culture. In the East, people struggled to sort through the war dead to count them, and sometimes to identify them by name or unit or side in the conflict. But the intent was to give the war dead proper burial. Out west, explorers, military men, and curious amateur ethnographers dug up the dead and sent body parts, most often skulls, to Washington, where individual identities disappeared as crania, sorted by tribe, were reduced to plain numbers. (By contrast, the identity of nearly every donor was noted in the records and published in the catalogs.)

Work on skulls had become more systematic too. In the 1860s, a generation of Americans learned the importance of counting and measuring, as the country mobilized for war. They operated in a world becoming "numerical and measured in every corner of its being," to borrow an apt phrase from philosopher Ian Hacking. Draft boards identified and sorted recruits, measured them, and put them in uniform. The impulse to sort and measure applied to the dead as well as the living. Field commanders counted losses, as best they could, when they sent reports to superiors.[12] The effort to number the dead continued after the war, as men assigned to burial details counted dead comrades and settled their bodies into marked graves. Their work provides a striking contrast to that of the army doctors out west who dug up graves and sent remains east, where colleagues then counted and measured the dead, but arranged their body parts on museum shelves or stashed them in storage drawers.

While government officials named and numbered the war dead, while American communities from New England to the Ohio Valley settled soldiers' bodies into the ground and raised monuments on town greens, Native American communities in the West watched the dead disappear from cemeteries and burial sites. As historians have discovered, community work on the Confederate dead played into Southern nostalgia for the Lost Cause. But what happens to our picture of the nation and its dead when we include Native American remains in this history? We don't usually think of the war's dead and the Native American dead as pieces of the same story, but to look at how differently the dead were treated exposes a long root of the late twentieth-century's repatriation scandals. The mix of military might and intellectual curiosity that characterized the growth of nineteenth-century empires brought bodies into museums in Europe and the United States. But events in the American Civil War and practices that grew out of that war (particularly its new emphasis on America's dead) give an especially poignant character to imperial body collecting in the United States. Dead bodies cannot suffer, but communities of survivors can.

Professor McGill's letter reminds us of something every student of mourning rituals knows: though an individual may die, the rituals carried out around death and burial preserve the community, which endures. American collectors, some of whom helped consecrate burial grounds in the East, casually desecrated burial sites in the West. The work of collecting bodies and measuring them that had begun during the war continued in the West, but with very different consequences.[13] The story begins with efforts to establish an army medical museum, where McGill and his comrades deposited specimens of their wartime surgeries.

Burying the Dead

The Army Medical Museum had a slow start, but in time enthusiasm for the collection spread through the Medical Corps and donations arrived in Washington. A first catalog, published in January 1863, lists 988 "surgical preparations" (body parts) and 133 "extracted projectiles" (bullets). Curators cataloged gunshot fractures to the tibia, fibula, neck, radius, ulna, elbow joint, astragalus, sacrum, scapula, clavicle, humerus, head of the humerus, shaft of the humerus, hand, ankle, finger, carpus, wrist, and on through an unhappy anatomical litany. Early donations included 59 shoulder joints, 138 arms, 56 forearms, 14 hips, 436 thighs, 161 legs, and "the mutilated limbs of seven general officers." (Distinguishing an officer's leg from the leg of an ordinary soldier could be difficult, but labeling an officer's body part could encourage a "humbler" individual to donate "fragments of his own mutilated frame for this sacred purpose," Brinton believed.) In fall 1863 the army took over rooms from the Corcoran Gallery of Art and opened the collection to visitors, who were "attracted by a new sensation."[14]

Battlefield surgery was the collection's primary emphasis, but the museum's donors contributed a number of human skulls, a body part that better expressed mortality than surgical artistry. Surgeon W. H. Rulison of the Ninth New York Cavalry donated the cranium of "Private T.B., 'F,' 7th Michigan Cavalry, captured at Gettysburg, and cut down by a Rebel Lieutenant, because wearied, he fell behind on the march of 3rd July; admitted cavalry Corps Hospital, 4th July; died forty-two days, after injury, 15th August 1863." Rulison did not explain why he cut off the man's head, nor did Surgeon M. K. Hogan of the United States Volunteers, who sent a "metopic calvarium" (the top portion of a skull) of a soldier who committed suicide.[15] Others simply collected skulls left scattered on battlefields. The museum acknowledged a contribution of thirty-two skulls that New York Surgeon Reed Bontecou "picked up" on the battlefield of the Wilderness.[16]

Bontecou is best known for wartime medical photographs that document surgical procedures, but he must have been a strange sight gathering up skulls. One of the war's most famous photographs captures the chaos of body parts left by hasty battlefield burials. "A Burial Party, Cold Harbor, Virginia, April, 1865," a picture taken by John Reekie, appeared in Alexander Gardner's *Photographic Sketchbook of the Civil War*. It depicts "soldiers in the act of collecting the remains of their comrades" killed at Confederate victories at Gaines' Mill in June 1862 and Cold Harbor in June 1864. The bodies had been left to rot in the open air or tossed into trenches too shallow to hold them through a winter freeze and a spring thaw. (During spring months, the muddy ground of old battlefields sometimes coughed up skulls.) According to Gardner, the Virginians who lived near Cold Harbor had failed to fulfill a simple human obligation to bury the dead. The picture "speaks ill of the residents of that part of Virginia, that they allowed even the remains of those they considered enemies, to decay unnoticed where they fell. The soldiers, to whom commonly falls the task of burying the dead, may possibly have been called away before the task was completed. At such times the native dwellers of the neighborhood would usually come forward and provide sepulture for such as had been left uncovered." Concern for public health should have brought out battlefield neighbors armed with shovels and ready to work, but cash-strapped Virginians thought the Union should pay them to bury the dead.[17]

Instead of paying Virginians, the Union sent a burial detail of African American soldiers. One of these soldiers poses beside a stretcher filled with pieces of human bodies. He has seated himself on the ground to align his flesh-covered skull with the bare skulls of his dead comrades. In the background, four soldiers bend over shovels. The camera focuses on five skulls on the stretcher at the center of the picture, one for each of the living men at work among the dead. The picture's composition enhances its power. The human remains form an axis that separates the viewer from the workers but draws all into a central fact of common mortality. It is possible that these men on burial detail tried to distinguish Confederate from Union dead, taking clues from scraps of cotton or wool uniforms. In later years museum catalogers moved all these battlefield skulls into one "White" section, erasing both political and

Fig. 27. "A Burial Party, Cold Harbor, Virginia, April, 1865." Negative by John Reekie; print by Alexander Gardner.
(Courtesy of the Library of Congress.)

ethnic differences in an enlarged racial category. Things had changed since Morton's day, when a careful craniologist pulled out skull charts and checked to see if he could spot an African, Polish, Swedish, Irish, or German head.[18]

An old craniologist might also have remembered ornithologist John Kirk Townsend's comment that skull collectors benefited from the epidemic diseases that destroyed a community's ability to bury its dead. The Civil War proved a similar boon for those who collected body parts. At the war's end, the museum had specimens enough to fill six sections: the Surgical Section, the Medical Section, the Microscopical Section, the Anatomical Section, the Section of Comparative Anatomy, and the Section of Miscellaneous Articles (a boot, a crutch, a necklace of fingernails, and parts of rifles and muskets). Museum collections preserved more than 200,000 wounds, the results of more than 40,000 operations, and more than 6,000 separate specimens of military surgery, exhibiting "the effects of missiles of every variety on all parts of the body." Visitors could study excisions, fractures, "osseous specimens," "preserved dry, neatly cleaned, mounted on little black stands," wet preparations preserved in alcohol, 350 plaster casts "representing the mutilations resulting from injuries and surgical operations," and 400 "curious freaks of bullets," preserved to illustrate "the effects of percussion upon the missiles themselves."[19]

Visitor numbers got a boost in 1867 when the museum reopened in Ford's Theatre on Tenth Street, which had been closed since Lincoln's assassination in April 1865. Contemporaries commented on the building's strange new status, asking "what nobler monument could the nation erect to his memory than this somber treasure-house to the study of disease and injury, mutilation and death?" Those words came from Dr. J. J. Woodward, one of six doctors who had performed an autopsy on the murdered president. At noon on April 15, 1865, with their work lit by Washington's spring sun, they cut open the president's body. Woodward had the unhappy task of tracking the bullet through Lincoln's brain to make official "the facts of death by homicide." The memory of his work on that awful day may explain his wish to see a tribute in body parts.[20]

In the coming years, the nation did indeed build nobler monuments to Lincoln, but in the first years after the war, Washington sightseers who came to honor the martyred president often wandered up the theater stairs to the grisly collection. They passed a floor of busy pension clerks bent over records of the dead and wounded and glanced in at the chemists testing drugs in the Surgeon General's laboratory, and then discovered an upper story filled with the museum's "unique collection" of war materials. That close association with Lincoln meant that the Army Medical Museum displayed its specimens for an unusually large group of "non-professional persons of both sexes," whose interests were often sentimental and sensational. Postwar tourists came to see a display of John Wilkes Booth's brain, heart, and vertebrae as well as famous skulls, like the skull of cannibal chief "Vendovi." Popular magazines warned that the "timid would not care to visit at midnight and alone. Fancy the pale moonlight lighting up with a bluish tinge, the blanched skeleton and grinning skulls—the same moon that saw, in many a case, the death blow given or the bullet pierce."[21]

But the museum collection sometimes left non-professional visitors scratching their heads. "What is the use of it all? What good is to be expected from this laborious and painstaking collection of mutilated and diseased fragments of the human frame? Why should they be so carefully put away into bottles or locked up in cases, and such efforts made to secure their permanent preservation?"[22] The easy answer: medical education. But curators also invited sober-minded visitors to indulge a healthy curiosity about "the structure of their own bodies, the function of certain organs, the arrangements of parts in certain localities where they have felt pain or discomfort, or the changes which have caused death in relatives or friends." Curators reserved specimens illustrating the anatomy of genital organs for "qualified medical visitors" but created a section on human reproduction for general visitors. As interest in the wartime surgery faded, visitors lingered longest at an exhibit on the development of the human embryo. A "lady may go alone, and, unnoticed may learn at her leisure something about her own peculiar function."[23]

By the 1880s, the Army Medical Museum combined attributes of a war museum, an anatomical cabinet, and an ethnographic laboratory. Each year the museum attracted some forty thousand visitors, who had a chance to see a collection that merged anatomy and ethnography. The collection included remains from caves and tumuli in Greenland, Alaska, Florida, and Arizona; specimens from "the majority of the existing tribes of Indians, and of the extinct tribes of the historic period"; and "skulls of the white and black races." A doctor in Cincinnati sent nine tiny fetal skeletons; curators purchased wax models of eyes, ears, brains, muscles, and nerves produced at Vasseur's Paris workshops and by Baltimore doctor (and wax artist) Joseph Chiappi. Displays included a miscellany of body parts: "four pelves of Sioux Indian squaws," the pelvis of a Frenchman, bones of the left hand of an adult woman, the right hand of a male mulatto, a brain or two, "four hyoid of bones, of criminals who were hung," and the "dried hand and foot of a Piegan Indian, found tied to a stake on the prairie." But human skulls—more than three thousand of them whole and in pieces and cataloged as crania (all the bones of the head and face), calvarium (bones without the lower maxilla), and calvaria (bones of the skull alone)—dominated the collection.[24]

Despite the museum's varied functions, collections preserved a tie to Morton's cabinet of crania. In January 1873 Surgeon General Joseph K. Barnes and his museum curator U.S. Army Surgeon George A. Otis asked Congress for $5,000 to print a catalog of the *List of the Specimens in the Anatomical Section of the United States Army Medical Museum*. This small investment promised substantial returns in knowledge and scientific prestige. Like James Aitken Meigs, who cataloged Morton's collection, Otis and Barnes argued that students of craniology (including "foreign savants" in France and Germany) needed information on "large collections." "The medical officers of the Army have collected a much larger series of American skulls than have ever been available for study," they wrote, channeling Meigs's "hints" about cataloging skull collections. They acknowledged that Samuel George Morton had launched American craniology but complained that his *Crania Americana* had been available by subscription and only to a wealthy few.

They knew too that America's Civil War had dated Morton's work, changing the country's underlying assumptions about race, its ideas about statistical methods, and even its ideas about dead bodies. Morton's connections to phrenology embarrassed a new generation of scholars, and his arguments that species did not change seemed out-of-date to Darwin's followers. Otis and Barnes cataloged their museum's miscellany of body parts and sorted its skulls by race and gender, measuring facial angles, diameters, lengths, widths, and internal capacity, and presented it all in the numerical form "most convenient for the greater number of anatomists occupied with these studies."[25]

Anatomists looked first for the results of skull measurements, although those with a sense of history would have noted the nine named individuals whose skulls the museum had collected, measured, and cataloged: " 'Vendovi,' chief of one of the Fiji Islands"; " 'Wah-ker,' a celebrated chief" of the Pah Ute; " 'Hee-taw,' sub-chief of Bannock tribe"; " 'Weak Eyes,' one of Two Kettles' band"; " 'Shota,' chief of Ogallallas"; " 'Walk-a-bed,' " an Arapahoe medicine man; " 'Cunning Jim,' a Cheyenne chief"; and Comanches " 'Eath-Ath,' Red, 'Qua-ha' Day. Killed near Fort Richardson, Texas," and " 'Tooh-Parrah,' Black Bear" and his wife.

Their stories thread through the "Ethnographic Section" and remind us that military violence persisted through the century and kept up the steady supply of crania. The 1873 Modoc War on California's lava beds produced four crania, packed in a barrel and forwarded to Washington with a note from Assistant Surgeon Henry McElderry, who explained that these were the men executed for the "murder of Maj. General Canby and Rev'd Thomas, Peace commissioner. The heads are labeled with the respective names of each."[26] The Mormon militia attack on an emigrant group passing through Mountain Meadow in 1857 produced a child's calvarium; the 1864 "battle at Sand Creek, C.T." produced two Arapahoe crania.[27]

No doubt many crania in the army's collection have stories that twist through contingencies as complex as those that characterized Veidovi's history. The numbers convenient to anatomists may seem to hide those histories, yet there are stories behind efforts to find a mathematical language to describe human bodies. By the century's end, Americans seemed to operate in a world of "unlimited statistics, endless columns of figures, bottomless averages merely for the asking," as Henry Adams wrote.[28] Gathering that data made for strange encounters with the living and the dead.

Measuring the Living

During the forty-eight months of the American Civil War, the federal government put uniforms on 2,753,723 men. In the process of making men into soldiers, the government recruited draft board physicians to gather up a lot of information about 605,045 conscripts. Where were they born? How tall were they? How long were their arms? How big were their feet? How thick were their necks? Army doctors also took the measure of 225,639 men who enlisted voluntarily, 79,968 who were hired as substitutes, and added in data on 8,000 sailors, some students from Harvard and Yale, and a few hundred Native Americans. Data on health and on size provided information useful to military commanders, clothing manufacturers, insurance companies, and to the students of the great Belgian statistician Adolphe Quetelet, who were interested in describing the average man of every race, nation, and ethnic group.

It is no surprise that there were plenty of glitches in attempts to measure living men. Draftees faked defects; volunteers hid theirs; and no two doctors measured in exactly the same way. A perfect version of the encounter went like this: a naked draftee "was asked his name, age, nativity, and occupation, and questioned in regard to his general health and that of his family, whether any hereditary taints existed, and if he had ever suffered from any disease or accident, thus endeavoring to obtain all the information possible concerning him, his conversation at the same time enabling the surgeon to judge his mental as well as his physical qualifications." Subjective portion of the exam over, would-be soldiers were sent to be measured with a special apparatus devised by a

Dr. S. B. Buckley, who personally trained technicians to use his equipment. But the army's need for a fighting force trumped its need for accurate measurements, and Dr. Buckley complained that the military drafts left too few skilled mechanics free to build his devices.[29]

Even imprecise measurements could yield data for comparisons. "Single men," for example, appeared to be healthier than married men; "the men of dark were healthier than those of light complexion; . . . the youngest men, exclusive of those under suitable age or deformed, were healthiest; and that disease increased steadily with increase of age. . . . Of all nativities, the Indians of the United States were found most healthy, they being affected by but few of the diseases common among civilized men; and it may be proper to note here that the only diseases on account of which any of them were rejected are syphilis, scrofula, diseases of the eye, and chronic disease of the bone."[30]

The draft produced a portrait of the "average American" sliced in various ways. But just how did Native Americans come to appear in the statistical portrait of American soldiers? Data on Indians was notoriously hard to come by. As Canadian ethnographer Sir Daniel Wilson remarked in 1858, "It is no easy thing to obtain actual measurements of Indians' heads. I have seen an Indian not only resist every attempt that could be ventured on, backed by arguments of the most practical kind; but on solicitation being pressed too urgently, he trembled, and manifested the strongest signs of fear, not unaccompanied with anger, such as made a retreat prudent."

Twenty years later, geologist F. V. Hayden echoed the sentiment: "Those who have never attempted to secure photographs and measurements or other details of the physique of Indians, in short, any reliable statistics of individuals or bands, can hardly realize the obstacles to be overcome." It took "tact and perseverance" to overcome their superstitions and their seemingly "irrational" fears. Yet Hayden acknowledged an entirely rational basis for resistance when he noted that "usually it is only when an Indian is subjected to confinement that those measurements of his person which are suitable for anthropological purposes can be secured."[31]

The Civil War produced a few exceptions to patterns of Native American resistance. In western New York, Dr. Buckley found a cooperative group of "five hundred Indians, including all the full-grown men of unmixed race accessible on the Iroquois reservation." These Iroquois men were taller than white soldiers, except for those born in Kentucky and Tennessee, the "length of *head* and *neck* is small, like that of the Negro, averaging but 9.55 inches, or 0.4 less than for white soldiers." They had a greater "*breadth of pelvis*," a greater "*circumference of thorax*," broader faces, and longer feet than white soldiers. Their arms were "longer by more than an inch and a half on the average than that of the white."[32]

In 1862 a doctor in Green Bay measured 130 Wisconsin volunteers, most likely the Chippewa men who enlisted to help the U.S. government fight their old Sioux enemies in Minnesota. As a group, their health was pretty good; "only about ten per cent were rejected as physically disqualified, and most of these for extensive cicatrices from burns or incised wounds. Only one case of hernia occurred, but neither varicocele nor varicose veins of the extremities. There were three or four cases of scrofula and secondary syphilis." "I learn," he added, "from officers commanding these men that they were good soldiers, being unsurpassed for scouting or picket duty, but quite unable to stand a charge or artillery fire."[33]

The Chippewa helped capture a group of "uncivilized Indians" whose captivity set the statisticians' hearts to beating. A Sanitary Commission actuary who looked over the draft board data despaired of producing an objective portrait of Native Americans. The numbers were too few and taken by too few men to correct for errors and eccentricities. He was sorry to report that the War Department thwarted efforts to measure "the large number of full-blooded Indians, who were held for a considerable time as prisoners of war near Rock Island, in the upper Missouri." Authorities in Washington concluded that "the scientific results did not promise to be of sufficient value to warrant the introduction of irresponsible persons into our large prison camps."[34]

"Irresponsible" statisticians had their eyes on Sioux prisoners held at Camp McClellan in Iowa, 265 men that President Lincoln spared from a Minnesota gallows in December 1862. The prisoners had been convicted for their part in the "Sioux Uprising," an episode whose aftermath captures the peculiar character of skull collecting—the connections between histories of violence, war, death, and disruption, and the search for accurate methods to measure physical differences among human groups. In August 1862, Dakota warriors in Minnesota turned against white settlements in the southern part of the state, beginning the bloody battles of the long conflict between the United States and the Sioux that would not end until the Massacre at Wounded Knee in

1890. Treaties signed in the 1850s altered the relationship between the U.S. government and the Ojibway (or Chippewa) and the Dakota and a small number of Winnebago who lived on lands in the Minnesota Territory between the upper Mississippi and the Missouri rivers. In 1851 and again in 1858, the Dakota exchanged land for reservations and annuity payments. Some may have seen a promise of peace or prosperity in the treaties, but annuity payments made the Sioux dependent on the federal government and, closer to home, dependent on particularly corrupt and incompetent traders and agents.

Encounters grew violent in the summer of 1862. Annuity payments were late that year, and the ill-tempered local agent refused to distribute provisions on credit. "So far as I am concerned if they are hungry, let them eat grass," he is claimed to have said. On August 17, four young men took some eggs from a white farmer's henhouse. They argued over their prize, but then turned their anger on white settlers, killing three men, a woman, and a little girl, and launching an all-out war on Minnesota's frontier settlements.

More than 400 men, women, and children died in the next few weeks, but by mid-September white settlers, backed by soldiers, gained the upper hand. Some Dakota took off for Canada; others surrendered to Colonel Henry Sibley, who appointed military commissioners to handle Indian trials. Commissioners made quick work of 393 cases, convicting 323 men of murder and rape and sentencing 303 of them to be hanged. In the summer of 1862, Abraham Lincoln was not paying much attention to the Minnesota frontier. But when the convictions came down, he ordered the Justice Department to review each case. On the department's recommendations, the president spared the lives of all but 38 of the Sioux, permitting the execution of only those who had been found guilty of rape and murder. The rest, he decided, were enemy soldiers, and he sent them off to Fort McClellan, where they were protected from lynch mobs and measurers alike.

Settlers, measurers, and medical men, however, got their hands on the dead. An acting assistant surgeon sent to Washington a pelvis he was sure "belonged to one of the hostile Sioux tribes." "I am sorry to state that the body was badly mutilated by soldiers and citizens, before I was able to secure it; one of the hands and scalp having been cut off and carried away, the lower jaw being fractured from blows besides other unnecessary ill treatment of the same."[35]

The bodies of others fared better, if it is possible to think such a thing.

On December 26, 1862, 38 Dakota men were hanged in the largest mass execution in American history. The dead were buried in a single shallow grave, but local doctors, including William Mayo, drew lots for the bodies. Mayo's famous doctor sons claimed they learned their first lessons in anatomy with their father's prize—the skeleton of Cut Nose.[36]

Those spared the noose were sent with their families to Fort McClellan in Iowa. They spent the next four years as army prisoners. In 1866 President Andrew Johnson pardoned the Sioux who survived their captivity and sent them to the Santee Reservation near Niobara, Nebraska. Their fertile Minnesota reservations had been claimed by white farmers—happy to take advantage of the recently passed Homestead Act. With the Sioux removed, Minnesota's future looked promising to whites, or so it must have seemed to those who read the *New York Times*. Where the "Rebel Sioux" had begun a "frightful holocaust of blood and flame which swept over the Minnesota frontier last Autumn, there grow this Spring flowers, peace, promise, and lasting safety."[37]

Still some eastern commentators wondered whether Minnesota settlers had lost their moral bearings in the quest for peace and safety. Territorial governor Alexander Ramsey offered a bounty for scalps, "just as other states pay for wolf scalps," reported a sympathetic observer. "It is a 'State Right' that will not be given up." The raw brutality of the governor's offer caused a scandal, and Ramsey began to pay cash instead to those presenting " 'satisfactory affidavits' of their exploits in bringing down their human (or inhuman) game." In other words, what seemed brutal and uncivilized was not so much hunting and killing but actually presenting rotting bits of skin and hair to clerks in a governor's office. Words on paper were a more civilized kind of evidence to carry to the governor.[38]

In July 1863, a Minnesota farmer killed Dakota leader Little Crow, who had fled to Canada rather than surrender. The farmer discovered Little Crow and his son picking berries and shot them both. He collected his $500 bounty from the state. Whether he presented a scalp or an affidavit, we do not know. Nor do we know whether he kept the head of his celebrated catch. Evidence gathered from the correspondence of bone collectors suggests that he might have considered doing so. And that he might have passed his prize along to sons and grandsons, and that each, in his turn, told the story of the skull. Others followed McGill's wartime lead and donated prize skulls to the Army Medical Museum, which made room on its shelves for an "ethnological series." But that name hides the smaller wars that fed the series through the 1870s and 1880s. In many ways, despite its ethnographic turn, the Army Medical Museum remained a war museum.

"Ethnological Series"

The Army Medical Museum began as a wartime project, but the Civil War's political distinctions and the identities it created as men joined opposing armies settled into the background when skull collecting turned back to biology and ethnography in the postwar years. In April 1867, Surgeon General Barnes sent "a circular letter" to medical officers in the field and "invited" them to help the museum by collecting "1. Rare pathological specimens from animals, including monstrosities. 2. Typical crania of Indian tribes, specimens of their arms, dress, implements, rare articles of their diet, medicines, etc. 3. Specimens of poisonous insects and reptiles, and of their effects on animals." A quick response came from Acting Assistant Surgeon R. B. Hitz, who sent the museum a Crow Indian war club, a set of elk antlers, and "two Indian Crania (perfect) which I exhumed myself. The one is a 'Spokane' and the other a 'Pagan' skull."[39]

Barnes needled Hitz's slower comrades, asking post commanders to urge upon "medical officers in your department the importance of collecting for the Army Medical Museum." He explained that the "progress of anthropological science" depended on a large supply of crania. "There are many Acting Asst. Surgeons who would doubtless collect such things if they knew they were desired. I have thought it better to attain the object by correspondence, rather than by circular order, which by making the task obligatory might make it distasteful." Medical men "have already enriched the Mortonian and other magnificent craniological cabinets by their contributions, and it is hoped that they will evince even greater zeal in collecting for their own Museum." The Surgeon General knew that some who missed the chance to collect body parts during the war would be happy to send "donations of Indian crania, of specimens of natural history, and of objects of ethnological or archaeological interest."[40]

McGill's war work would have made him a likely collector on the plains, but there is no record that he received the April circular before he left New York. McGill headed west in the summer, hitting the plains before the weather had turned cold enough to tempt body collectors back to work. But that fall, doctor colleagues, many working on contract for the army in the West, responded to Barnes's circular and began to send crania to Washington. These doctor-collectors have gotten bad press from recent scholars who see an intellectual callousness and a racist arrogance in the apparent ease with which they dug up native bodies and shipped remains to Washington. Collectors left some nasty descriptions of their grave robbing, especially when they snatched bodies in front of grieving relatives and boiled flesh off fresh corpses.[41] And it is hard to miss a contrast in the fact that just as one federal agency spent money to rebury bodies, another paid freight on bodies unburied in the West and shipped east.

Even though they often noticed that people disapproved of their grave robbing, few doctor-collectors seem to have considered that craniology carried a potential to harm the living. Many army surgeons pictured themselves holding a shaky middle ground between land-hungry settlers and the Indians whose lands settlers coveted, and most seem to have imagined that work on skulls would add to the store of ethnographic knowledge.[42] In the late twentieth century, their work brewed into the scandals that helped push through legislation on repatriation and reburial. But for a handful of army doctors, skull collecting seemed a hobby to fill a frontier fort's long and empty hours. Superiors in Washington worried that doctors grew bored during peaceful and healthy months, when no soldiers came by complaining about diarrhea or conjunctivitis, when there were no bullet wounds or snake bites, no fights or accidents, and no pregnant laundresses. The Surgeon General recommended hobbies to break up long days: "woman's knitting work," or sketching, photography, botany, entomology, herpetology, Indian languages, or craniology.[43]

Taking Washington's advice, Assistant Surgeon George Kober set aside his experiments on beakers of urine to devote "some of my leisure hours . . . to the exploration of Indian burial grounds, and the collection of crania and skeletons." He found quick rewards in shallow graves, where he discovered fascinating specimens of desert-dried mummies.[44] Dr. James Kimball reassured his father that "with my books, a few congenial persons, a gun and something to shoot, I can be contented anywhere." He added skull collecting to his pleasant pastimes and contributed Assiniboin, Yankton, and Blackfeet crania to the museum.[45] Thanks to men like Kober and Kimball, that first season's fieldwork quickly produced forty-seven native skulls for the museum: Tsuktshi,

Flathead, Chenook, Selpish, Nisqually, California, Piegan, Spokane, Mandan, Arickaree, Gros Ventre, Sioux, Kaw, Minataree, Menominee, Cheyenne, Kiowa, Arapahoe, Wichita, Navajo, and Apache crania.

These donations had a curious effect on the skulls back in Washington. They seem to have prompted curators to create the new "Whites" section for their war skulls, a decision that fit well with new cultural patterns that underlined differences between white Americans and "others," particularly Native Americans and African Americans. Museum catalogers fell in with a general consensus that pulled immigrant groups from Ireland and Germany into the "white race." The war's generous helping of dead bodies also solved craniology's perennial shortage of "normal" white skulls, as dead white soldiers were drafted to represent their race, assuming the role that Morton had had to assign to low-lifes and criminals, like that man-eater Pierce. Yet the new category could be troubling. At least one student of craniology in Britain wondered at an influx of soldiers' skulls from battlefields in Crimea. Skulls of eighty-seven soldiers represented "Natives of Britain" in the Museum of the Army Medical Department at Fort Pitt. He concluded that British soldiers were "a class of men who are frequently wholly uneducated, and their mental faculties undeveloped, and are therefore not the best specimens of British skulls." Americans, on the other hand, accepted the war's windfall, pleased to have dead soldiers represent the white race.[46]

The late 1860s and early 1870s were heady times in the museum's new quarters in Ford's Theatre; bad years for Native Americans, but banner years for collectors. Forty-seven skulls gave the ethnological collection a good start, but they were too few to provide cranial statistics on any tribe or group. In September 1868, the Surgeon General sent out a new "Memorandum," encouraging medical staff to contribute specimens needed to promote "the progress of anthropological science by obtaining measurements of a large number of skulls of the aboriginal races of North America." "While exotic and normal and abnormal crania of all descriptions are valued at the Museum for purposes of comparison, it is chiefly desired to procure sufficiently large series of adult crania of the principal Indian tribes to furnish accurate average measurements." Museum staff turned from work on war wounds and diseases of the army and embraced a statistical project to "illustrate the morphological basis of ethnological classification, more especially of the native races of America, including anthropometry and craniology."[47]

As John Shaw Billings explained, "The measurement and comparison of human skulls is an interesting and difficult branch of the natural history of man . . . that can only be pursued by a few anatomists, having access to large collections of crania, and the requisite experience and skill to measure accurately." Even a collection of three thousand skulls was "not actually half large enough to permit drawing conclusions from it."[48]

The project tripped over its uncertain intellectual foundations. Did racial differences actually appear as differences in skull shape or size? Did the cranial morphology of tribal groups actually differ? And how many skulls would it take to make a case for anthropometry? During these years, European craniologists advised collectors that fifty skulls could establish the traits of a population from a burial ground. American collectors, still puzzled by the variety of aboriginal populations, seem to have taken all the skulls they could find, compiling samples from groups distinguished by culture, but not by biology. Craniologists also ignored histories of captive taking and lovemaking that should have led them to question assumptions about racial or biological differences. For example, to create categories for skulls of Brule Sioux, Ogallala Sioux, Yankton Sioux, Sisseton Sioux, Santee Sioux, Teton Sioux, and Wahpeton Sioux was more a collector's boast than a biologist's fact.

Skull numbers could balloon with surprising ease, especially when collectors worked in the communities along the Pacific coast, where tens of thousands had died from epidemic diseases and from the violence of the Gold Rush years. By the 1880s, there were more skulls of native Californians than any other group. A collector described a single trench that yielded ten to fifteen tons of bones, a "*Big Bonanza*," he called it. He and his colleagues pulled bones out until the sun set. Curators counted up 459 skulls from the Channel Islands off the Santa Barbara coast. Only an "old crone [who] for many years continued to visit this spot annually to mourn the departed greatness of her people" disturbed the collector's view of the scene.[49]

Acting Assistant Surgeon George P. Hachenberg must have come pretty close to what the deskbound officers in Washington imagined as a model field collector. Hachenberg, a Pennsylvania-born surgeon who had served in the Union army, reenlisted in the late 1860s, hoping dry western air would clear up his congested lungs. The army sent him to Fort Randall, on the western banks of the Missouri River, just across the Nebraska border in present-day South Dakota. Life in the West suited polymath Hachenberg. Trained as a dentist, he set himself up as a daguerreotypist to pay his way through medical school at New York University, where he finished up with the thesis "Music as a Therapeutic Agent." Over the years he experimented with electricity and telegraphy; raised bees; and designed a vegetable cutter, an envelope, a hygrometer, a torpedo, a baking powder, a corn-weevil poison, and a burglarproof safe. He believed he could catch a man in a lie by taking his pulse, thought the Senate could register votes electronically, and looked for a way to pipe gas from coal mines to businesses in cities. Hachenberg was "somewhat eccentric in his mode of thought," a contemporary wrote. One scholar has called him an American Leonardo da Vinci, but that may be going too far.[50]

Hachenberg's eccentricity (and a taste for what Morton's Egyptologist friend George Gliddon called craniology's "rascally pleasure") made him a good skull collector. "I got so interested in my labor in writing the Medical History of the Post and more lately in my Skull hunting expeditions that I found it pleasant to stay longer," he wrote to the Surgeon General, who was pleased to learn that he had a staff member contented at his post. Over some four months in 1868–69, Hachenberg sent the museum a Winnebago cranium, seven Ponca skulls, "four pelves of Sioux Indian squaws," and more than fifteen Sioux crania. He wrote that skull collecting had taken him traveling "more than 300 miles sometimes on foot—sometimes on horseback—sometimes in the night, and sometimes in the day time," often in the company of a "fearless" tracker and guide named "Heck." (Did the museum have an extra $100 he could pass along to loyal Heck?)

Ethnographic collectors often pretended that the human remains they collected were principally of historic interest—relics of a vanished race, the remains of ancestors long forgotten by living descendants. Yet Hachenberg's accounts overflow with evidence to the contrary, and he seemed proud of his ability to outfox the living. He sent reports to the Surgeon General and to the eastern newspapers describing his exploits. He eyed the skull of man "whose father was a Yankton, and mother a Brulé." Relatives planned "to keep watch over the body," but Hachenberg and Heck "snatched his head, before he was cold in his grave" and before his kin had set their guard. The two climbed a scaffold and took a skull in broad daylight. He was "seen by a number of Indians; but after getting it, instead of going away from them, I joined them, leaving them under the flattering impression that curiosity led me to examine this imposing arrangement of their dead."[51]

Collecting satisfied Hachenberg's private fantasies (or maybe brought back his early dental training). Though he shipped most of the skulls he collected to the museum, he kept the skull of a young Ogallala woman because she had "remarkably beautiful teeth . . . every tooth was perfect and of the most symmetrical order." When the Dakota air yellowed the teeth, he parted with the skull, which became specimen "**2034.** Cranium, F. æt. C. 35, Cap. 1200 c.c., L. 171 mm., B. 132 mm., H. 128 mm., I.f. m. 44, L.a. 338 mm., C. 485 mm., Z.d. 130 mm., F.a. 67°. From near Ft. Randall, Dakota, Territory. Presented by G. P. Hachenberg, M.D."[52]

Or skull collecting simply made for good stories. Picture him, skull in hand, surrounded by Fort Randall comrades, sipping whiskey, smoking cigars. He holds up the "cranium of a young squaw that was the mistress of Lieutenant Long, who was on duty at this post, before the War of the Rebellion. At the breaking out of the War he abandoned her, and went south, where it is said he was killed in one of the battles of the War. The squaw having remarkable beauty after he left her, she captivated a private soldier, who kept her tenderly up to the time of her death. She was burried [*sic*] in a coffin, by her last lover in the river bottom land, between the Fort and the river. She died of an abscess, located on the side of her neck. What its nature was I could not learn. I searched many days, and dug as many holes in finding the remains of this subject, and when finding it, was not disappointed in securing a fine specimen." She becomes one of fourteen Yankton Sioux, "**499.** Cranium. F. æt. c. 20, Cap. 1285 c.c., L. 173 mm., B. 135 mm., H. 123 mm., I. F. m. 44, L.a. 349 mm., C. 489 mm., Z.d. 126 mm., F.a. 70°. From near Fort Randall, D.T. Presented by Acting Assistant Surgeon G.P. Hachenberg."[53]

In the 1860s, a philosophical observer hoped that the Civil War specimens could be used to help doctors learn battlefield surgery. Surgical lessons saved lives. In the 1880s, an observer could watch museum workers translate skulls into statistics, not medical lessons. Skull measurers went to work on the army's large collection, challenged to find ways to measure this strange corner of the world. They knew enough about statistics to know they needed good samples and reliable techniques to produce accurate average measurements. But the search for representative samples remained challenging.[54]

Measuring the Dead

Chance, violence, and enthusiasm of men like Dr. Hachenberg helped fill out the museum collections, but how could the Army Medical Museum make the case that its collection was any more representative than Morton's cabinet had been? Curators turned to the market, purchasing anonymous "Negro" skulls from Richmond, Virginia, and anonymous "White" skulls from Professor William Gibson's "Anatomical Cabinet." Receipts in the Army Medical Museum files record price skulls between $3 and $5. (Damaged heads sold for as little as $2.50.) In the 1880s, curators paid $5 for a fine French skull, but $3 apiece for skulls from Paris cemeteries, and $3 for the "skull of a Congo Negro; face below the eyes missing." Three dollars also bought the "calvarium of the celebrated Rojas, ferocious chief of Guerillas, killed by a soldier of Capt. Berthelin's Co., French Army, time of Maximillian." Anthropologists contributed too. In the 1890s, Franz Boas acknowledged that peddling skulls was "unpleasant work" but helped finance his studies.[55]

The museum competed with collectors like British craniologist Joseph Barnard Davis, who published a *Crania Britannica* (which opened with a decorative medallion of the overlapping profiles of Morton and Blumenbach, "the father and grandfather" of the science) and a *Thesaurus Craniorum*. Davis and his colleague John Thurman studied crania hoping that the form of skulls contained keys to a history of the ancient inhabitants of the British Isles. Donors sent him skulls they dug from cairns and ditches, but Davis also purchased skulls, including the contents of a museum of comparative anatomy, as well as skulls from English graveyards, Spanish prisons, and French cemeteries. He was especially pleased with Dayak skulls from Borneo, where "a good skull is regarded to be worth as much as a slave, *i.e.* about £25 in our money."[56]

In the 1880s, the museum relied on Mr. George Kiefer to fill out its South American collections. In June 1888, Kiefer sent Billings eighty-eight skulls ($3 each), along with pots and books and pamphlets he purchased for the Surgeon General's library. He passed along what he had learned, describing catgut strings as a good way to fasten lower jawbones to the upper half of a skull. He had seen skulls made into masks too. But if the museum wanted to know more, he would have to he paid. "I[,] like all antiquarians, naturalists & poets[,] am always poor," Kiefer wrote. Collecting was dangerous. "The mountain specimens are very difficult to get and require a great deal of hazardous climbing[;] should you see the grounds you will wonder how without the aid of wings or balloons the Incas managed to reach such localities to bury their dead."

Kiefer's work brought him to a sad end. An 1889 auction catalog for the final sale of his collection begins with this note:

> The collection of Peruvian Antiquities was gathered in person by the late George W. Kiefer, Esq., who devoted nine years of arduous labor to this work and spent large sums of money in the undertaking.
>
> To the cost in time and money must be added that of life itself, for, owing to the irritating dust arising from freshly opened graves, Mr. Kiefer contracted lung desease [*sic*] to which he succumbed a few months ago, in this city, where he had come to rest and with the hope of disposing of his Collection entire to some of our Museums.

Bidding for the museum, Billings bought two Peruvian mummies for $20 each, another for $18, and six human skulls for $3 each. Three dollars is about what he would have paid a photographer to take his picture.[57]

Back in their labs, craniologists were not worrying about stirring up "irritating dust from freshly opened graves." They needed to figure out how to measure the skulls and how to describe and publicize their work. Morton had used pepper seed and buckshot, borrowed terms from phrenologists like Combe and ideas from Egyptologists like Gliddon. He had hired artists to draw skulls onto lithography stones. Techniques had changed in the decades since Morton's death. The contrasts can be striking. Late nineteenth-century craniologists used water to gauge the internal capacity of skulls and composite photographs to capture differences in their shape and dimension. They took up statistician Adolphe Quetelet's search for the average measurement for any human group, pursued Francis Galton's quest to find a means to measure anything and everything, and adopted terms devised by French anatomist and anthropologist Paul Broca and his disciple Paul Topinard.

In 1875 Broca published *Instructions craniologiques et craniométriques*, a new guide for craniologists. We remembered Broca for his discovery of the area of the frontal lobe where speech originates. But he also developed methods for measuring skulls. He apologized that the volume had taken longer than he expected, but he had measured more than 1,500 skulls and isolated 200 significant cranial elements. Before craniologists could decide significant questions about cranial variation, they had to agree on how to measure skulls. Broca offered specific instructions. Fifty skulls were sufficient to establish a statistical profile of the population of a given ossuary, but Broca recognized that collectors would sometimes take as many skulls as they could find, falling for the collector's excess that tipped into controversy a century later. Broca did not anticipate that Americans would collect as many skulls as they did. Look for complete skulls, he said, and skulls with a jawbone. Brush off the dirt. Make notes about a skull on the forehead: sex, age, race, tribe, and the place where it was found. Glue in loose teeth. Fasten the jawbone to the zygomatic arch with strong cord, and wrap each skull in a separate package in case pieces rattle loose in shipment.[58]

Broca recommended measuring the internal volume of skulls with lead shot, but he thought any regular granular substance or some liquid could be just as effective. The irregular shape of skulls made accurate measurements difficult, and it was hard to get consistent results from different measurers. In the 1880s, craniologists at the Army Medical Museum tried to figure out how to capture differences among skulls. They presented two papers at the National Academy of Sciences on April 22, 1885. Curator John Shaw Billings commented briefly with *On Composite Photography as Applied to Craniology*, and his museum colleague Washington Matthews (who collected Sioux skulls in the Dakotas but is better known today for his work on the Navaho) described his work "On Measuring the Cubic Capacity of Skulls."[59]

In the 1880s, measurers like Francis Galton, an eccentric polymath credited with promoting the science of eugenics, and craniologists at the Army Medical Museum began to work with composite images, pushing the visual fad for pictures created from multiple exposures of a single negative. Composite photography provided "a rapid and convenient means of obtaining a graphic representation of a series of irregular objects, a picture that would indicate the mean and size and shape of these objects, but also to a certain extent, the maxima and minima of their variations." Galton made composites of Roman women, Greek queens, Napoleon, Cleopatra, and Alexander the Great, using old coins and marble busts. He tried the technique on live criminals, consumptives, and Jewish schoolboys. The closer the resemblance among members of a group—the tighter the genetic links—the better the composite image. It was hard to line up the faces of criminals, consumptives, and asylum inmates, for example. And to add to the challenge, one of Galton's mad subjects, who hated waiting for his chance to pose, rushed around the camera and bit the stooping photographer on his backside.[60]

Photographers had better luck with students at Smith College, gentlemen members of the "The Monday Evening Club," a "Family of Eight—Father, Mother, Five Boys and Girl," Omaha men and Dakota women, and "American Men of Science," although the scientific logic behind the groupings can be hard to find. Anthropologist Alice Fletcher, who posed the young Dakota women, thought the blended images confirmed what she had learned during her years of fieldwork. Composites captured "a people, intellectual rather than brutal, unawakened rather than degraded." And a confident friend thought the composite of scientists conveyed "an idea of perfect equilibrium, of marked intelligence, and, what must be inseparable from the latter in a scientific investigator, of imaginativeness."[61]

These happy thoughts sprouted on thin scientific ground. The technique promised to be more effective with inanimate objects, like skulls. Museum workers have left some eerie images of skull composites made from crania of Sandwich Islanders, Esquimaux, Sioux, Ponca, Arapahoe, Ogallala, Piegan, Minataree, Apache, Comanche, Wishitaw, and Pah-Ute. For a visually challenged statistician, here were "equivalents of those large statistical tables whose totals, divided by the number of cases and entered in the bottom line, are the averages." But the images captured the edges of those averages—the "maxima" and "minima" of cranial variation—in strange, shaky lines. With their blurred outlines, the beautifully printed images read like illustrations of a ghost story, the unburied dead come back, teeth chattering, to haunt naughty children.[62]

A second presentation at that April meeting brought the audience back to earth with old questions about how to measure the internal capacity of skulls. Morton had set an American standard with "granular substances," like buckshot and seeds, and a handful of followers had tried to use beans. Their efforts produced nothing more than rough averages, often failing the simplest scientific test of repeat results. The differences among skulls were often minute, and subjective expectations skewed results when measurers, little conscious of their actions, packed a few more seeds into a skull they supposed had once housed a bigger brain. Subjective errors were magnified because "laws regu-

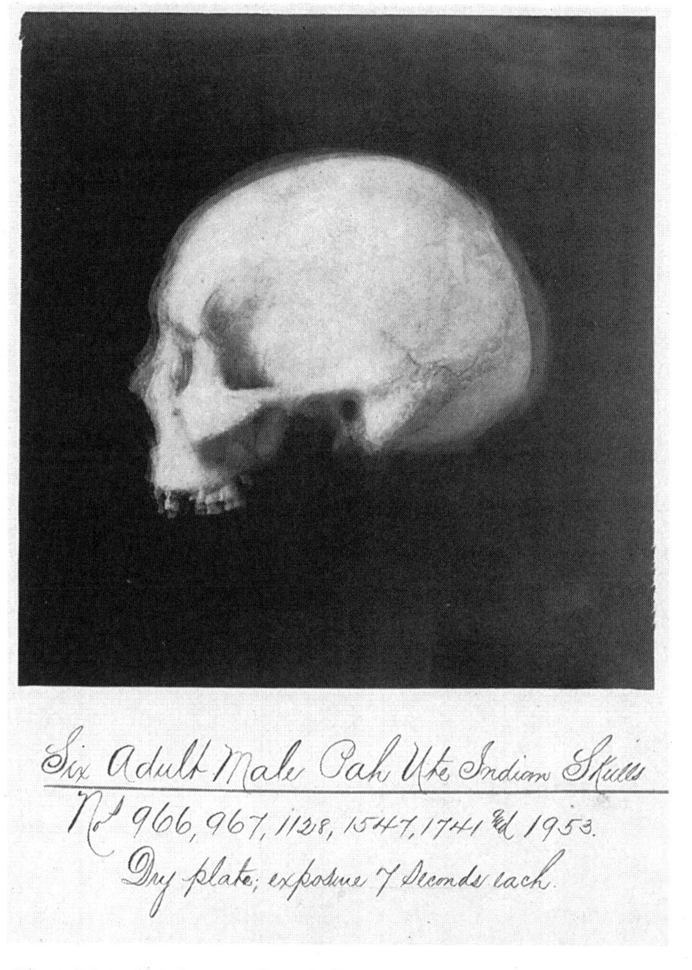

Fig. 28 "Six Adult Male Pah Ute Indian Skulls" (ca. 1885–87). War Department, Surgeon General's Office, United States Army Medical Museum, Washington, D.C. (Photography Collection, Miriam and Ira D. Wallach Division of Art, Prints and Photographs, The New York Public Library, Astor, Lenox and Tilden Foundations.)

lating the fall and subsidence of granular substances are imperfectly understood."

Until physicists worked out those laws, Matthews recommended that skull measurers use water. There are "no sciences more widely or well understood" than "hydrostatics and hydraulics," he said, and "the perfect knowledge we possess of the laws which govern its motions

Fig. 29. "Ascertaining Capacity of Cranial Cavity by Means of Water" (1884). War Department, Surgeon General's Office, United States Army Medical Museum, Washington, D.C.
(Photography Collection, Miriam and Ira D. Wallach Division of Art, Prints and Photographs, The New York Public Library, Astor, Lenox and Tilden Foundations.)

would be of vast advantage" in cubing skulls. Water promised objectivity, since there was no way a scientist with "muscular exertion" and a yen to make one skull bigger than another could pack more water into a skull. True, water soaked into porous bone, settled into "squamous sutures," and leaked out the sinuses, but a rubber coating solved those problems. According to Matthews, who filled a skull with water and set a metronome to ticking off the seconds, with this new method of measuring "the unchangeable element of time takes the place of the fickle element of vital force" and gave craniologists a chance to get an accurate gauge on skulls. Matthews pushed for objectivity, but even by the standards of the 1880s, skull measuring was not an exact science. Matthews's assistants sometimes stuffed too much putty into empty eye sockets, sometimes too little, and their sloppy work slowed an already inefficient process.[63]

When Matthews published his lecture, he added illustrations that could help settle the curiosity of those puzzling over the fate of skulls collected out on the plains. Think of the watchers who saw Dr. Hachenberg and his tracker Heck snatch skulls from burial scaffolds. "What is the use of it all?" they might have asked, along with other Army Medical Museum visitors. Here they could see Matthews and a colleague at work among the collected skulls, emptying water from one, removing putty from another, setting one on a scale, posing another in the case of a craniophore, the apparatus for making a composite photograph. Who could have imagined the remains of dead friends and relatives in such an odd scene? The woven basket was a familiar object. But the basket throws the picture off. Apparently, the modern lab used a very old technology to fetch and carry skulls. The presence of the basket-bound pair of skulls also has the effect of turning the photograph into another accidental memento mori, like Gardner's picture "A Burial Party, Cold Harbor, Virginia, April, 1865." "What is the use of it all?" the skulls ask as they grin up at the busy pair working at their lab tables. The search for precision and certainty is only vanity.

Puzzling relatives had another chance to catch a glimpse of the remains of their collected kinsmen when the Army Medical Museum contributed twelve skulls, a few composite photographs, and a craniophore to the United States display at the Columbian Historical Exposition at Madrid in 1892. The labels on the case read:

> Skull of a Nisqually Indian chief, Puget Sound, Washington. The flattening is extraordinary.

> Skull of a Peel River Indian, Fort McPherson, Arctic America (Jukkuth-kutchin). From Mr. R. Kennicott's collection.

> Skull of a Pawnee Indian, near Fort Harker, Kansas. Presented by Dr. B.E. Fryer, surgeon, U.S.A.

Skull of an Arapahoe Indian warrior, from Fort Larned, Kansas. Presented by W.H. Forwood, assistant surgeon, U.S.A.

Skull of a Ponka Indian, from Fort Randall, Dakota. Presented by Dr. A.J. Comfort, assistant surgeon, U.S.A.

Skull of a Piegan Indian, of the Blackfeet Nation. Killed near Fort Shan, Montana. Presented by Dr. F.L. Jown, surgeon, U.S.A.

Skull of a California Indian, from Santa Rosa Island, California. From Rev. Stephen Bower's Collection.

Skull of a Brulé Sioux Indian from Beaver Creek, Nebraska, 4 miles north of Camp Sheridan, Nebraska. Presented by W.H. Corbusier, assistant surgeon, U.S.A.

Skull of a Wahpeton Sioux Indian from near Fort Sisseton, Dakota. Presented by Mr. A. Gecks, hospital steward, U.S.A.

Skull of a Nez Percé Indian, from Bear Paw Mountain, Montana. Presented by Dr. David Snively, assistant surgeon, U.S.A.

Skull of an Eskimo of Alaska, from the northwestern extremity of St. Lawrence Island, Bering Sea. From Mr. E.W. Nelson's collection.

Skull of an Alaskan Eskimo, from the northwestern extremity of St. Lawrence Island, Bering Sea. From Mr. E.W. Nelson's collection.[64]

What did visitors make of the display, which paired anonymous Indian heads with their named American sponsors? Historians suggest that Spain sponsored the exhibition to boost its national prestige, which had been battered in Europe and was threatened by the United States. The United States joined the twenty-four countries participating in displays, pageants, and lectures that paid homage to Columbus, to Western civilization, to Spain, and to the accomplishments of scientists and collectors. Organizers awarded the museum's exhibit a silver medal, just as they had to Morton's forty-four skulls. The 1893 World's Columbian Exposition at Chicago would help turn Columbus into an American hero and launch the United States into the twentieth century; those silent skulls on display at Madrid seemed to lock Native Americans into a dead past.[65]

Curators knew that history was not so simple. If they listened to their chatty collectors, they learned that the specimens they sent to Madrid carried traces of a very recent human past. Dr. Snively had sent the skull of a Nez Percé man killed in a fight with U.S. troops in 1877 at Bear Paw Mountain in the Montana Territory. Or consider the Pawnee skull from Fort Harker. Dr. Fryer sent a note with the skull.

> I had already obtained for the Museum the skull of one of the Pawnees killed in the fight you speak of and would have had all had it not been that immediately after the engagement, the Indians lurked about their dead and watched them so closely, that the guide I sent out was unable to secure but one. Until within a day or two the snow has prevented a further attempt. Yesterday I sent a scout who knows the spot and think I can get at least two more of the crania—that number being reported to me as left unburied by the Pawnee, and it may be that if the remaining five (eight not seven were killed) are buried or have been hid near where the fight took place—about twenty mile from here—I can, after a time go obtain all. I shall certainly use every effort.[66]

"Walker's War"

It is obvious from these accounts that skull collecting remained an odd practice, that even if it gave rascally pleasure to collectors, it disturbed the people whose relatives were collected. In the late 1870s, however, a few scholars pressed through the violence and sensation to explore the information on mortuary customs and burial practices that turned up when collectors dug up the dead. Civil War veteran Dr. Henry C. Yarrow wrote *Introduction to the Study of Mortuary Customs among the North American Indians* (1880) and *A Further Contribution to the Study of the Mortuary Customs of the North American Indians* (1881), early publications of the Bureau of American Ethnology that began to take culture practices seriously. Like Billings, McGill, and other doctor-collectors, Yarrow reenlisted in the army after the war. He nursed soldier colleagues through the cholera epidemics that hit postwar army bases in Georgia and New York, and then headed west in 1871, serving as surgeon and naturalist on army Lieutenant George Wheeler's geographical survey of arid lands west of the 100th meridian. The desert's snakes and lizards fascinated Yarrow. (Herpetologists know that the beautiful blue-green spiny *Sceloporus jarrovi* is named for Yarrow.) During his years in the West, he also stuffed hundreds of bird specimens for the Smithsonian and pulled skulls from graves, mostly in California and Utah. Army Medical Museum catalogs thank Yarrow for ninety-three skulls from Santa Barbara and at least seven Pah-Ute skulls.[67]

Yarrow's intellectual history is difficult to trace, but he and bureau director John Wesley Powell seem to have brought a different sensibility to the study of graves. Strains from America's hard years of mourning and measuring echo through Yarrow's work. But so does a new interest in the idea of culture: a growing conviction that culture, not biology, makes men human. We can sense the change in Yarrow's turn from collecting skulls to studying mortuary customs. When Powell introduced Yarrow's first study, he described the subject's importance. Mortuary customs reveal a people's philosophy, expressing their beliefs about the "nature of human existence in life and after death, and the relations of the living to the dead."

At Powell's prompting, Yarrow sent a circular to Indian agents, amateur ethnologists, surgeons, and army officers, asking them to gather information on "Manner of burial, ancient and modern," "Funeral ceremonies," and "Mourning observances, if any." Yarrow wanted a book that covered all forms of "burial," a word he directed contributors to take from its old Anglo-Saxon root—*birgan*—"to conceal or hide away," by inhumation, cremation, embalmment, aerial sepulture, or aquatic burial. True that environment sometimes dictated burial practices, but Yarrow hoped his study would reveal cultural links among groups of people. "These materials constitute something more than a record of quaint customs and abhorrent rites in which morbid curiosity may revel." "The mystery which broods over the abbey and where lie the bones of a king and bishop, gathers over the ossuary where lie the bones of chief and shaman; for the same longing to solve the mysteries of life and death, the same yearning for a future life, the same awe of powers more than human, exist alike in the mind of the savage and the sage."[68]

Yarrow knew he could count on a generation of observers trained during years when death and burial topped the nation's cultural agenda. Articles in popular periodicals took up some of the issues in the anthropology of burial customs to help create a consensus that the federal government should assume the obligation to rebury the war dead. "The usual custom, from time immemorial, has every where been to commit the dead to the bosom of mother earth," an author wrote, as he urged readers to encourage the United States government to bury the war dead.[69] While reburials continued in the East, Yarrow excavated burials in the West. California trenches yielded hundreds of anonymous native skulls. But a "cairn or rock burial" in Utah attracted particular attention. Yarrow dug up the remains of " 'Wah-ker,' a celebrated chief," one of the few skulls in the army's collection with an individual identity.

Yarrow described his excavation of Walkara's burial in his *Introduction to the Study of Mortuary Customs among the North American Indians*. His tone is flat and objective. "Several of the graves were opened and found to have been constructed in the following manner": at the foot of a rock slide, in a cairn, lined with animal skins. The site, he wrote, was hidden on the side of "an almost inaccessible mountain" and would "have been almost impossible to find without a guide." "The corpse placed therein, with weapons, ornaments, etc., and covered over with saplings of the mountain aspen; on top of these the removed bowdlers were piled, forming a large cairn, which appeared large enough to have marked the last resting place of an elephant. In the immediate vicinity of the graves were scattered the osseous remains of a number of horses which had been sacrificed no doubt during the funeral ceremonies." Yarrow reported that he found the body of a chief in this elephantine grave. He also "found parts of the skeleton of a boy, and tradition states that a captive was buried alive at this place." He gives neither a date nor a source for that tradition, adding only that ancient inhabitants of the Balearic Islands in the Mediterranean had also buried their dead beneath heaped stones. Yarrow excavated what was, in fact, a recent burial.[70]

Walkara lived an eventful four decades. For a time, he was one of the most powerful men in southern Utah, deploying skills as a diplomat and horse thief to lead his people through a tricky world of white contact, as Mexico and then, after the war in 1846, the United States claimed to govern the territory he had dominated. In the 1830s, Walkara teamed up with American fur trappers to raid Mexican ranches in southern California. Local legend holds that he once stole three thousand horses, earning him a reputation as "the greatest horse thief in history." He traded people too, capturing Paiute and Digger women and children to sell to New Mexican slave traders. In some accounts, he provided women for sex-starved American mountain men. When the Mormons arrived in 1847, Walkara, with his eye on the future of the horse trade, threw in his lot with Brigham Young and invited Mormon settlers to farm Ute lands in central Utah. In 1850 a Mormon bishop baptized him a "brother-in-the-faith." But relations with the Americans soured quickly, particularly when Mormon settlers monopolized the region's fertile streams and hunting grounds. In 1853 Walkara led his people in a series of raids that Mormons dubbed the "Walker War" or Walkara's War.

Walkara's War began with a small transaction gone wrong. Mormon annals report that in the summer of 1853 a Ute woman came to a Mormon settler's cabin with trout to trade for flour, a commodity scarce in Ute households that year. Ute families were poor and hungry in the early 1850s, and many had suffered and died of measles in the winter of 1852–53. The two women dickered over a fair trade and, when they couldn't agree, called in their husbands. The settler husband offered three pints of flour for three trout. Three pints seemed too little to the Ute husband, who turned his anger over the bad bargain on his wife. He began to beat her, the Mormon report continues, until the settler intervened and hit the man over the head with a gun barrel. The Ute man died a few hours later. The small dispute laid bare deeper problems between the two communities. Mormons took the moral high ground, certain their compatriot had been right to stop a wife beating and certain they were also in the right to try to stop Walkara's trade in captives and slaves. Walkara complained on behalf of his people that Mormons settlers had taken Ute lands and that "the graves of their fathers had been torn up by whites."[71]

Walkara asked that the Mormon settler (farming Ute lands by Walkara's permission) be surrendered and tried in a council of Utes. The Mormons refused. Rumor spread through Mormon settlements that Walkara had found allies among the Apache and Comanche and that together they had "sworn a war of extermination" against the white people. "Just give me twenty-five, fifty or a hundred men, and I will go and fetch you Walker's head," a settler promised Young. All through the summer and fall of 1853, the Utes raided Mormon farms.[72]

Walkara and the Utes lost the war, a last series of battles in efforts to control patterns of settlement on their lands. Walkara pushed for a treaty that would insure Ute control over Utah lands, but the movements of people and animals and the trade, war, and diplomacy that had once given him opportunity for power had settled into a new pattern. When daguerreotypist Solomon Nunes Carvalho met Walkara in

1854, he described him as the "king of an aggrieved and much injured people," once a man of imposing appearance now careworn and haggard.[73] Walkara died of pneumonia in January 1855, and he was buried with elaborate ceremony. Mormon histories report that followers and kinsmen buried Walkara's body among the boulders in that elephant-sized cairn. They say the Utes tossed a boy and a girl into the burial pit to honor Walkara by dying beside him. They say passersby refused to free the boy, who complained that Walkara's rotting corpse had begun to stink. Did this boy with the unhappy death become "(**968.**) Cranium. Child æt. c. 6, Cap. 1070 c.c., L. 164 mm., B. 132 mm., H. 114 mm., I.f. m. 44, L.a. 323 mm., C. 460 mm., Z.d. 102 mm., F.a. 83°. Presented by Acting Assistant Surgeon H. C. Yarrow"?[74]

In letters to museum curator George Otis and Surgeon General Charles H. Crane, Yarrow explored the site's history and touted his own bravery in leading an expedition to recover a skull he knew belonged to "Wah-ker, a celebrated Ute Chief, long the terror of the people of Utah, New Mexico, and California." Walkara had been dead almost twenty years. Yarrow bribed and cajoled his way to the burial site and sent the museum Walkara's skull and "the cranial bones of a Piede or Piute Indian, said to have been buried alive with him." He explained that the "Chief was born about the year 1815, on Spanish Fork River, Utah, and died at Dry Creek, near Kanosh, Utah, in 1858. After his death the tribe carried the body up a mountain in the vicinity of their camp, and after killing a number of horses and two prisoners, buried it in a rock slide, from which place the writer . . . removed it. Your attention is called to the fact that in the inferior maxillary bone of the chief, the 'wisdom teeth,' so called, had never appeared above the margin of the alveolar process."

The skull becomes "(**966.**) Cranium. M. æt. c. 45, Cap. 1350 c.c., L. 180 mm., B. 134 mm., H. 132 mm., I.f. m. 39, L.a. 360 mm., C. 506 mm., Z.d. 140 mm., F.a. 76°. From Dry Creek, Utah. Presented by Acting Assistant Surgeon H.C. Yarrow. ('Wah-ker,' a celebrated chief.)"[75] The buried captive becomes "(**967.**) Calvaria, imperfect. From Dry Creek, Utah. Presented by Acting Assistant Surgeon H.C. Yarrow." Too imperfect to measure, but too interesting to leave behind. But chief and captive both appear in the composite photograph that the museum staff produced. It is a ghostly image, yet the numbers tattooed on the skulls are clearly visible, keys to the specific life histories and complicated relations of this man and this child.

Fig. 30. "Six Adult Male Pah Ute Indian Skulls" (ca. 1885–87). War Department, Surgeon General's Office, United States Army Medical Museum, Washington, D.C. (Photography Collection, Miriam and Ira D. Wallach Division of Art, Prints and Photographs, The New York Public Library, Astor, Lenox and Tilden Foundations.)

When Professor McGill wrote to John Shaw Billings, he had begun to have doubts about America's ability to remember his son. Emigrants, farmers, and railroad men, each of them working hard to advance America's future in the West, conspired to erase any sign that George had died and been buried on the Colorado plains. In the exceptional care given to Walkara's burial, Yarrow witnessed a similar concern about the need to carry memories of the past into the future. The accidental conspiracy that obliterated George's grave was played out around Native American graves as intentional policy, particularly during a few brief years in the late 1860s and early 1870s when this phase of skull collecting was at its height. In Yarrow's case, curiosity and scientific interest, good questions generated by archaeologists and physical anthropologists, played into an unsettled politics around Utah land claims.

Did Walkara's enemies lead Yarrow to the chief's grave? Did local people play on the curiosity of craniologists to settle an old score? Or had a new generation of Ute leaders surrendered control over the land where Walkara was buried in the hopes of establishing stable relations with American neighbors? I am not sure we will ever find a precise account of events, which occur, after all, at grave sites when the story's principal actors are dead and gone. Curators who cataloged body parts and measured skulls wrote epilogues to life stories, mixing insult with tribute, but preserving traces of Walkara, just as their entries on Dr. McGill's donations preserved traces of the life and work of Professor McGill's heroic boy.

The difficulties in gathering the requisite material, and even the crude data alone, have been and still are very great; in fact they are sometimes insurmountable. Religious beliefs, sentimentality and superstition, as well as love, nearly everywhere invest the bodies of the dead with sacredness or awe which no stranger is willingly permitted to disturb. It is seldom appreciated that the remains would be dealt with and guarded with the utmost care, and be used only for the most worthy ends, including the benefit of the living. The mind of the friends sees only annoyance and sacrilege, or fears to offend the spirits of the departed.

—ALEŠ HRDLIČKA, *Physical Anthropology* (1919)

EPILOGUE

Brains, Bones, and Graves

The Brains of Eminent Men

Readers leafing through copies of the *New York Times* on September 29, 1912, might have paused at a full-page feature with this arresting headline: "Dissecting the Brains of 100 Famous Men for Science." Print columns surrounded a portrait of Dr. E. Anthony Spitzka, well-turned-out professor of anatomy at Philadelphia's Jefferson Medical College and "prosector" of the American Anthropometric Association. The story promised that Spitzka had found that "actual weight and tissue of the brain are significantly correlated with mental superiority." Good news again for big-headed people. But Spitzka had a plea for the paper's readers. If anyone had a famous and accomplished relative about to die, let prosector Spitzka know. His association would be happy to claim the brain, since the "special object" of his society was the "promotion of scientific knowledge regarding the organization of the human brain in persons of superior intellectual capacity, as compared with those of the lower classes and also different races."

Spitzka and brain-measuring colleagues in Philadelphia and at Cornell represented the American side of a movement that had begun at the Society of Mutual Autopsy of Paris in 1881, an association founded with the "express purpose of securing elite brains for scientific study." The time had come for scientists to find specimens to represent themselves. For a few years, the Civil War had eased the shortage of white skull specimens, but brain collecting added a new dimension to the preservation and display of human remains. Spitzka and his colleagues encouraged families to imagine that it was an honor, not an insult, to have a relative whose brain had been deemed worthy of collection. Some who preferred to keep corpses whole balked at their efforts. But not sophisticated initiates who donated their brains to serve science. And they could consider their brain donations without stumbling into the cultural obstacles that had made it impossible for Morton's family and friends to take his skull and set it among his specimens. What did it mean to preserve and display a brain? Did a human brain conjure death and defeat? Or did it recall the butcher's counter, last week's good dinner of calf's brain? Brains did not work like skulls. They rotted too quickly to serve in the cultural arsenal that underscored an enemy's humiliation. There were few symbolic associations to haunt new brain displays. No horror-filled memories of rotting heads spiked on posts. No painted reminders of mortality.

Still, it is surprising that to display organs of consciousness could be so easily read as a compliment. Brainwork was cutting-edge science that demanded the best specimens, not just fresh brains but brains from the best and bravest men, men ready to buck convention and turn their bodies and minds over to science.

In 1906 Spitzka could describe eminent brains floating in jars in Paris, Munich, and Göttingen. He thought the Cornell Brain Association had about "seventy brains of educated, orderly persons." And his society had eight brains stored in a fireproof vault. Three had belonged to the society's founders. Unfortunately, a careless assistant had "flattened" two of those. The rest were in fine shape, including the brains of the anthropologist John Wesley Powell and the paleontologist Edward Drinker Cope, who in 1897 had been among the first Americans to will his body to science.[1]

The Anthropometric Association targeted the brains of celebrated men, although they needed a few ordinary brains to be able to distinguish the great. Spitzka said he had brains from dozens of disorderly and illiterate people, including brains of three brothers executed on the same day for murder. He had personally assisted at twenty hangings and sixty-eight electrocutions, and examined those "heads after death." Whether or not he preserved those brains, he did not say. Race slipped into the background as he tried to correlate genius and success with the size and shape of brains. White men became his favorite subjects.[2] But some anthropologist colleagues collected and preserved brains of Native Americans, including the brain of Ishi, the Yahi man found near a slaughterhouse in Oroville, California, in 1911. Press and public made a celebrity of "The Wild Man of Deer Creek," the "last surviving member" of his tribe, a man "untouched by civilization." Ishi lived the last years of his life in the University of California's anthropology museum in San Francisco, an informant and friend to university anthropologists. When Ishi died of tuberculosis in 1916, doctors cremated his body but sent his brain to Washington, an act that seemed to defy his cultural expectations and his explicit wishes. That exceptional brain must have held its own among the brains of eminent men, although its collection now seems wrong. What did scientists learn from it? Did they find race in this brain? Or just scars from old memories laid down during his years hiding from white Californians?[3]

The decades around the turn of the twentieth century saw the heyday of brain collecting. Spitzka and his colleagues believed that the brain's physiology explained why some men and some races were more successful than others. Like Morton, they tripped over their mistaken measurements and mistaken assumptions. Their confidence that brain size mattered, which linked them to Morton's old skull work, led them to odd speculations about successful men with tiny brains. Maybe age had withered their brains; maybe the brain-pickling solution had soaked into the folds and light liquid molecules had replaced heavy nerve cells. It's easy to laugh at some of their pretensions, particularly as neuroscientists recognize that weight is far too simple a measure for something as subtle and complicated as the workings of a human brain or for something as hard to chart as the connections between a brain's morphology and a man's or woman's social and intellectual success.

To turn to them at the end of this study of skulls brings us back to a history of the questions that drove Morton's skull work. Consider the three threads—a history of ideas about race, a history of human burial, and a history of American science—that have tangled their way through this story of skull collecting. These threads appear again as twentieth-century Americans continued to collect skulls, think about them, and argue over the fate of collected remains. No doubt, the preeminent scientific skull collector in the early twentieth-century United States was Aleš Hrdlička, the "father" of American physical anthropology. By the time Hrdlička died in 1943, he had collected and cataloged nearly twenty thousand human skulls, a number to dwarf poor Morton's grandest dreams. Hrdlička was a modern collector, ambitious in his collecting and efficient in his catalogs. His search for rare ancient specimens mixed with more recent burials prompted him to gather up as many skulls as he could find. For four decades, he added specimens to his bulging collection. As the twentieth century drew to a close, shock at the scale of Hrdlička's collection helped prompt a new conversation about burial rights, as Native American activists pushed Congress to pass legislation that mandated the return and reburial of Native American dead.

Another story surfaces in the life and work of Dr. Robert Wilson Shufeldt. The lure of race science continued to draw racist amateurs to Morton's old work of measuring skulls and sorting them by size into a hierarchy of races. Shufeldt was a doctor, a military officer, one of America's experts on the skeletal structure of birds, and, by a brief and unhappy marriage, a grandnephew-in-law of John James Audubon. In the 1870s and 1880s, he collected Native American skulls. Skull work stands behind two books that Shufeldt wrote to defend those who lynched African American men.

I know that bringing new characters into an epilogue violates every convention of a well-plotted book. But if you have come this far, you will know that I have tried to explore the personal histories and social circumstances that gave rise to ideas about dead bodies, race, and science. Although the actors may be new, the epilogue follows the ideas that have run through the episodes in this book. Why was it once fitting to take and keep these skulls? What purposes have they served?

"The Dead Man's Daddy"

Aleš Hrdlička, born in Bohemia in 1869, moved to upstate New York with his parents in 1882. Medical school launched Hrdlička's career in science, but he followed an unconventional path. He graduated from New York City's Eclectic Medical College in 1892 and two years later from New York Homeopathic Medical College. He went to work as a researcher at the State Homeopathic Hospital for the Insane and began to measure human bodies, joining the growing circle of anthropometrists. Although he wielded calipers among the living, Hrdlička took to skulls with unusual zeal and spent his career constructing and curating the most important collections of human crania in the United States. In 1903 he joined the new Division of Physical Anthropology at Washington's National Museum of Natural History as an assistant curator. He was appointed curator in 1910 and held that post until he retired in 1941, after a long and prolific career. In some four hundred publications, he wrote about human origins and evolution, speculated about the date of human settlement in the Americas, and promoted his new field of physical anthropology.[4]

Hrdlička died four decades into the twentieth century, but he struggled with Morton's intellectual legacy. In 1911 he answered a letter from Edward Nolan, librarian at Philadelphia's Academy of Natural Sciences. Nolan was puzzling over Morton's crania, and he turned to the country's best-known physical anthropologist to ask about the value of the sixty-year-old legacy. Nolan must have been disappointed to read Hrdlička's reply. Morton had assembled a great collection, Hrdlička acknowledged, but unfortunately his "measurements and observations are of only very little value today. He started, as you know, on the premises of phrenology, which late in the century had to be abandoned as entirely groundless." As he wrote elsewhere, "Morton may be justly and with pride termed the father of American anthropology," although a father "who left many friends to the science and even followers, but no real progeny, no disciples who would continue his work as their special of life vocation."[5] Brain man Spitzka gave phrenology a gentler nod, accepting its primitive insights into cranial location, but Hrdlička had other ideas, preferring to name France's Paul Broca as his discipline's "principal founder."

He adopted Broca's conventions on skull measurement, but not his recommendations on a sample size for skulls taken from a burial trench. Hrdlička was not certain statistical methods could help him solve puzzles about the history of human migration into the Americas. Quetelet's "Law of Errors" taught him that he would make as many mistakes measuring 250 skulls as he would make measuring the same skull 250 times, and if he wanted only a physical portrait of one group, he might have followed Quetelet and, as Broca advised, taken just a few skulls from a burial site. But Hrdlička had learned that "men of recent times have inhabited many of the sites that may have been occupied by early man, and it will be readily appreciated that human remains of different periods might often be closely associated or even intermingled."[6]

What if a burial trench contained one or two human skulls far older than all the rest? What if these remains held a key to the mystery of the origins of human populations in the Americas? Hrdlička's questions, like the intermingled bones at a burial site, left a troubled legacy for physical anthropology. Parties to recent debates about repatriation and reburial have sometimes stumbled over the confusions in collections like Hrdlička's.

In the late 1890s, Hrdlička oversaw the transfer of thousands of skulls from the Army Medical Museum's ethnographic collection to his Division of Physical Anthropology. Over the next forty years, he expanded his division's holdings with thousands of skeletal specimens he had collected himself. While some scholars describe Hrdlička as a careless man in the field (he collected too quickly to take notes, failed the archaeologist's test of describing soil layers), he was very careful with his fragile specimens. He asked his carpenters to build "tiers of sliding shelves or drawers" to protect bones from sunlight, moisture, rats, mice, and bone-eating beetles that threatened to return his prizes to their dusty beginnings. He taught assistants to boil bones just long enough to get rid of old flesh. (And don't let them steep too long in chemical baths, he added.) He sorted bones by age, sex, and race, and tagged each body or body part, noting "color, sex, age, nationality (if American, the nationality of parents in addition) and last disease; also the principal (not necessarily the last) occupation of the individual in his life." He measured the bones, hoping the data held clues to histories of evolution and migration, but he never escaped physical typologies of race. He put cleaned skulls into special drawers, separated normal bones from the odd and diseased, and devised a card catalog that gave him quick access to his whole collection.[7]

Contemporaries describe Hrdlička as a strange man, formal, idiosyncratic, and frenetic, dressed in a black suit, sporting a clip-on tie, and openly opposed to women in science. Writings leave us a difficult figure, who was sometimes deaf to complaints from those he measured and blind to contributions from native people he studied. Even eulogies and tributary volumes come up a little short on warmth. But no one questions his dedication to physical anthropology or his unshakable faith that disciplined scientific inquiry, as he understood it, would help assure a better human future.

A sense of his peculiar ways comes through in a diary he kept during his fieldwork in Alaska between 1926 and 1938. Hrdlička first headed north in the summer of 1926 to trace migration routes into the Americas, but spent enough time with the Aleuts to earn the name "Ashaalixnamaataax," the "dead man's daddy." "All the natives here already call me 'the skull Doctor,'" he wrote, with a pleased boast. Hrdlička was not known for great empathy, but he recognized that affection made collecting human remains a challenge for physical anthropologists. "Religious beliefs, sentimentality and superstition, as well as love, nearly everywhere invest the bodies of the dead with sacredness or awe which no stranger is willingly permitted to disturb." (Low pay was the other great challenge.)[8]

In Alaska, Hrdlička discovered that aboveground burials, customary in a frozen country, simplified bone collecting. The Arctic's long summer days added hours to his fieldwork. But bugs and people sometimes made the work challenging. Though his eyes swelled shut from insect bites, he laughed off a warning from a native woman that gnats and mosquitoes waited to attack grave robbers. And he tried to ignore the hard stares of locals as he packed up boatloads of skeletal remains. He remembered two native boys happy to help. He gave them empty sacks and sent them off to an old burial ground "to pick up every skull and jaw they can find. They go cheerfully," Hrdlička wrote. He went cheerfully to another site and packed 183 skulls into sixteen sacks. If he hadn't run short on canvas bags, he could have taken dozens more. He prized ancient remains, hoping they would help him trace human migrations, but contemporary dead fell into his large net. The 1918 influenza epidemic had killed so many Native Alaskans that government agents had had to bury the dead. Broken custom made it easy to pick up human remains. Ornithologist John Townsend described devastation like this for Morton when he wrote from Oregon in the 1830s.

Alaska proved a fantastic bone yard for Hrdlička, even though on occasion Alaska could disappoint. At one burial place, he found remnants of hundreds of skulls and skeletons but "so damaged as not to be worth saving." The damage was only partly nature's work. "The destruction was due to sailors and other whites who used to use the skulls for targets, and to dogs, reindeer and foxes." It's easy to imagine irreverent sailors taking pot shots at skulls, but it couldn't have been easy for local people to watch the sailors' games.[9]

Non-native locals helped Hrdlička too, especially Laura Jones, the wife of a cannery superintendent at Larsen Bay, who dug up dead workers, many from China, and shipped the skeletons to Washington, along with a brocade silk shirt, a jade bracelet, and a "pig tail." Though few women have appeared in this history, it's not hard to fit Hrdlička's Alaska friend into the annals of skull collecting. She was infected by the rascally pleasure that runs through collectors' accounts. Opening graves spiced up her life, just as it had spiced things up for George Gliddon in Egypt and George Hachenberg in South Dakota. Hrdlička, a man of science from Washington, affirmed her interest with his official stamp, something Morton's correspondents appreciated when they tried to imagine an intellectual life as they sweated through hot days on Indiana farms or in Florida swamps. It felt good when Philadelphia's naturalists, comfortable in stuffed chairs at their learned societies, sat up and paid attention to an idea hatched out on the frontier.

Picture Mrs. Jones with her shovel. She and her husband may have superintended their cannery workers when they were alive, but there seems to have been no community to assure these cannery men a good burial once they had died. Back at the museum, Hrdlička arranged Mrs. Jones's contributions among the remains of hundreds of others who had died far from home. Although the location of a death was not one of the categories Hrdlička used to sort his skeletons, he might have noticed that stories behind many of those whose bodies wound up in collections led back through chapters in the eighteenth and nineteenth century's linked histories of imperial expansion and scientific curiosity. (Those whose memories were filled with stories of a darker cast have followed collected bodies back to the annals of slavery, to memories of slaveholders who terrified bondsmen and women into submission by taking the heads of rebel kinsmen, posting them on stakes, and denying surviving relatives the right to bury their dead.)[10]

A Washington visitor, remembering these histories, might have looked at Mrs. Jones's skulls and recalled the Angolan boy, with his new teeth, who had died in Peter Camper's Amsterdam in the 1770s. Or Morton's anxious friend Dr. Doornik, who left Batavia in the 1820s, packing the skulls of a Dutch official dead from drink and a man from China hanged for forgery. Doornik examined a world that had scattered bureaucrats and thieves but shattered the communities that looked after their corpses. Doornik's own wandering life ended in New Orleans in 1837. Did friends come to mourn for Doornik, buy him a crypt? It doesn't look that way. The Chinook man William Brooks and the Fijian Ro Veidovi died far from home too. Brooks fell into the hands of a doctor who had a reason to bury his body and keep it buried. With different staff on duty at Bellevue Hospital that day, Brooks could easily have wound up as Ro-Veidovi's cabinet mate in the Washington museum, a young Christian's prized flattened skull to set beside the head of a cannibal chief.[11]

Mrs. Jones's Asian skulls let us tease yet another thread from the skull story. Could these workers have been Japanese or Filipino? Hrdlička asked, acknowledging that Asian and Pacific men had come to work in Alaska's canneries. He could see that these were not skulls of recently dead Alaska natives. Perhaps physical anthropologists could learn something from the skulls of contemporary Asian migrants about the ancient humans Hrdlička believed had followed game across frozen northern straits and into the Americas. Hrdlička's proposition satisfied a generation of physical anthropologists and archaeologists. It's the idea behind the Beringa hypothesis that many of us learned in grade school. But what if Hrdlička was wrong? What if there were multiple migrations to the Americas, as recent work suggests? What if humans headed out in small boats and settled along the coasts farther south in the Americas and far earlier than the Beringa hypothesis allows?[12]

Ironically, Hrdlička's collections could contain clues to a history that differs from the one he proposed. Yet his collecting work will make it difficult to know for sure. What we have come to see as the excesses of nineteenth- and twentieth-century skull collecting have made it difficult to use those collections to answer twenty-first-century questions. This is a case in which histories of ethics and politics have not lined up with the history of science. Hrdlička must have known this might happen when he acknowledged that superstition, religion, sentiment, even love, made people behave in particular ways around the dead. His search for answers to an ancient puzzle made it easy to ignore the sacredness and awe people had invested in corpses. But like those grave-dwelling gnats that attacked collector Hrdlička, sacredness and awe have come back to bite his intellectual descendants.

"The Boy with the 'Bent'"

A raw version of the racism that slavery's defenders extracted from Morton's work reappears in the writings of doctor-naturalist Robert Wilson Shufeldt. And the embarrassment of that racism also lingers in collections of skulls. Shufeldt was an expert on the osteology and paleontology of birds and a prolific author. His writings on creatures—

horned toads, Gila monsters, opossums, armadillos, manatees, peccaries, gophers, prairie dogs, bats, insects, and birds of all sorts (except ostriches and kiwis)—appeared in popular science publications at the turn of the twentieth century. According to one count, he had published more than eleven hundred articles by 1913. He might have been remembered as a curious enthusiast among a generation of nature writers, but he painted himself into an unhappy corner, using his naturalist skills—sharp observation, careful description, and adventurous collecting—to back misbegotten notions about racial difference. Naturalist credentials had helped Morton make a convincing case for his ideas about white racial superiority. Shufeldt, the naturalist, appears as a twentieth-century racist crank.

Shufeldt was a strange man. He was born December 1, 1850, in Manhattan and grew up in New York and Stamford, Connecticut, where the waters of Long Island Sound had enough game to keep an outdoor boy happy. His father—a naval officer, diplomat, and merchant seaman who served as Lincoln's consul general at Havana—took the family to Cuba at the start of the Civil War. The island's riches developed Shufeldt's passion for nature study. Trees full of lizards and frogs, moths that swarmed into a boy's room at night, waves teeming with small fishes, birds everywhere, and insects captured and pinned in rows in a cigar box—paradise for a boy with a bent for nature study. When Shufeldt's father rejoined the U.S. Navy in 1863, he took the boy along as a captain's clerk and signal officer aboard the U.S. gunboat *Proteus*.

That ship's discipline must have come hard to the thirteen-year-old, and nostalgia for the collecting boy creeps into Shufeldt's *Chapters on the Natural History of the United States*. He was pleased that courses in modern biology had opened new routes to serious nature study, but he remembered that "things were very different thirty-five or more years ago in my boyhood time, when a youth who gave evidence of any such tastes was commonly considered to be some sort of a juvenile crank, with a dash of insanity in his composition, and his father was advised to force him into one of the prescribed 'professions,' and make every effort to eliminate the eminently unpractical streak in his organization." He remembered a microscope taken away, the "rubbish" of birds' nests, eggs, and lizard skins ordered destroyed, living specimens made to be let go, and "the boy with the 'bent' bent sure enough into channels for which he had no taste or capacity." Even when his father indulged his collecting, Shufeldt recalled that he was disappointed that his firstborn son had no taste for a career in the navy.[13]

Shufeldt's currents ran sure enough back to naturalist channels, but eventually followed those channels to an extreme racist position. After the war, he studied medicine at Cornell and George Washington University and worked as a clerk at the Army Medical Museum, assigned to describe several thousand gunshot wounds to the humerus.[14] That task completed, he served as a surgeon in the frontier armies of generals George Crook and Philip Sheridan. He boasted that he was one of the first white men to visit the battlefield at Little Big Horn in June 1877, and despite a military detail sent to bury Custer's year-old dead, Shufeldt collected the skull of a trumpeter from Major Marcus Reno's command. (He donated it to the Army Medical Museum.) He spent five years in the 1880s at Fort Wingate, New Mexico, where he opened graves and collected Zuni, Apache, and Navajo skulls. Some went to museums in Europe and the United States. Others he kept.

The zest of the boy naturalist comes through in Shufeldt's "Personal Adventures of a Human Skull Collector," his recollection of 1910. Like most of those who collected dead bodies, Shufeldt boasted that he had outgrown the narrow superstitions and childish sentiments that stunted the intellectual lives of lesser men. (He described superstition and sentiment as particular weaknesses of African Americans and Native Americans.) Shufeldt's passion for crania began with a medical school skull, the head of a "mulatto man" hanged for killing a bank cashier. He had kept it all these years. The man had had a sorry life, but there had been no adventure for Shufeldt in the skull's collection.

He had better stories to tell about rare Navajo skulls. A young man who knew of Shufeldt's taste for skulls and was familiar with the rugged canyons around Fort Wingate brought him the skull of a man killed by a gunshot and the skull of a little girl. (Maybe she was six, Shufeldt thought.) Shufeldt tried measuring these skulls with water (as Washington Matthews recommended) but decided he preferred fine "dust shot," pellets smaller than common twelve-gauge shot. He photographed the skulls with his "excellent camera" and then traced the images onto lithographic plates. He sent the man's skull off to the Anatomy Museum at the University of Edinburgh, the museum where Morton had developed his interest in comparative anatomy.

The skull of another Navajo woman sat as a trophy in Shufeldt's study. Her story capped off his skull-collecting adventures. He told his readers that his "old washerwoman, a squaw known as 'Washie,'" had got herself caught up in drunken brawl. He watched as she was shot by a man who accused her of lacing his whiskey with kerosene. Before her corpse was cold, he began to plot to collect her body. He knew that by Navajo custom her relatives would pull the hogan where she had died down around the body. "Here was my chance," he said, and he took his sons along, hoping to collect a whole skeleton. But the woman's friends were waiting for them and shot at Shufeldt and his boys. They ducked behind the hogan's logs that night but returned for the body when the friends had dropped their guard. "At this writing the specimen is up on my mantelpiece in a place of honor. A large piece of scalp clings to the occipital region that adds a bit of additional interest to the specimen." He made another skull into an "excellent pen-rack."[15]

Shufeldt's skull trophies, rude though they are, appear less violent than the ideas he crafted about racial difference from African American skulls. He peppered his boyhood recollections with easy objectifications of the African and Chinese people he met, particularly when he sailed through the Caribbean with his father. African faces, "Mongolian" faces (that was the word he used), colored his world. Literally. They parade, dance, and drum. Play victim for his pranks (a poisonous frog dropped down a washerwoman's back) or cure an infected cat bite, concocting a remedy out of a solution of rotten eggs and mysterious herbs.[16]

In the early twentieth century, at the nadir of post–Civil War race relations, this naturalist dug his way back to some of the unhappiest conclusions of Morton's work—about the connections between the size of a skull and the size of a brain and about the low fertility rates of offspring born to parents of different races. Each of his two books, *The Negro: A Menace to American Civilization* (1907) and *America's Greatest Problem: The Negro* (1915), opens with an image of Shufeldt's medical-school skull resting on a pair of books and then explores the role of cranial morphology in "the more or less sudden stunting of the Ethiopian intellect." In Shufeldt's view, violence was inevitable when races mix. A reader with any doubt about Shufeldt's harsh views had only to look at the illustrations—photographs of the torture and lynching of Henry Smith in Paris, Texas, in February 1893—to understand the implications of his views on racial difference.

It's impossible to say why Shufeldt's work took this particularly ugly path away from nature study, away from his descriptions of the delicate bones of birds. His story about "Washie" suggests a man whose emotional arrogance might have stunted his intellectual development. Critics attacked his work, just as Douglass had attacked Morton. He sent his book *America's Greatest Problem* to anthropologist Franz Boas, who replied, "I am afraid that you would not wish to include an expression of my opinion in your book because I am not at all convinced that the miscegenation of the races is a bad thing from a biological point of view." If mixed-race children fall short of "the mark," Boas wrote, it was because contemporary Americans forced them to "grow up under very unfavorable circumstances." Shufeldt's racist turn pulled him from the mainstream of naturalist collecting, and contemporary reviewers shunned his book's "violence and prejudice," but his military service assured him a proper grave at Arlington National Cemetery, where he was buried in 1934.[17]

Reburying the Dead

Shufeldt's well-kept grave introduces a last thread: recent controversies over dead bodies. The body of the skull-collecting naturalist-racist lies in America's memorial to its military dead, a fate that differs starkly from that reserved for the Native American dead who were stored just across the Potomac River in Hrdlička's special drawers. Some of the contemporary issues take us back to the social inequalities that have made it possible to collect or preserve or display the remains of certain human bodies and to the disdain of collectors who ignored the wishes of the communities of those whose remains they collected.

In the 1970s, the injustice behind these collections of remains struck a cultural nerve. Native Americans, Native Hawaiians, and Native Alaskans who had objected to the double standard that upheld laws against robbing white graves but left the native dead vulnerable to collection and study finally found political traction in Washington. Why then? The horrors and struggles of the twentieth century and some of the plain silliness of that century's popular cultural attractions stand behind the shift in attitudes about skull collections. The clear links between race science and genocidal violence, the anti-colonial movements in Africa and the Pacific, and the work of activists who expanded the sense of civil rights pulled the collected dead out of museums in Europe and the United States. Tourist sites closed too, as activists underlined the irreverence and insult behind roadside displays of skeletal remains and open burial pits in Illinois, Kentucky, and Kansas. Roadside attractions that had been luring motorists since the 1930s began to close in the 1970s.

Still the struggle over the unburied dead has been a strange one. It is not easy to pry things loose from museums, even if visitors now shudder with a sense that a display of the dead "is not fitting." In some cases, the possibility that scientists working with technologies yet unknown will find ways to extract material from the collected dead either about diseases and diets, about present human differences or a common human past, has troubled efforts at repatriation and reburial. And however strong the recognition of past wrongs, there has been no simple consensus among tribes about what to do with the old bodies. Do they stay where they have been for the last one hundred years? Are the identities certain? Or are the bones of old enemies mixed with the remains of ancestors? Or do collaborations among native and non-native scholars, archaeologists, scientists, and curators promise new areas of knowledge and inquiry?[18]

On November 16, 1990, President George H. W. Bush signed the Native American Graves Protection and Repatriation Act (NAGPRA). That law, which built on a long history of state and federal legislation and followed a separate agreement with the Smithsonian Institution, protected graves on tribal and federal lands, prohibited traffic in Native American ancestral remains, and required museums and other institutions that received federal funds to publish inventories of the native remains and grave goods in their collections and arrange for their repatriation.[19] The United States joined a worldwide move for the return and reburial of bodies stored up from almost two centuries of collecting human remains. Responding to indigenous activists, European museums dismantled displays that smacked of colonialism's underlying racist assumptions and shipped remains back to America, Africa, and Australia.

The distinct history of American race relations and the scale of U.S. collections have colored American conversations over remains. Even Hrdlička's special drawers came into play. In 1986 a Northern Cheyenne man who had come to the Smithsonian's National Museum of Natural History to listen to a song from the Sun Dance stored in the museum's collection of wax cylinder recordings, asked about Hrdlička's walls of bone cabinets. Politicians were shocked to learn that they contained more than eighteen thousand specimens of human remains.[20]

Native activists, archaeologists, anthropologists, and museum administrators all weighed in on the law. Some worried about justice and human rights; others about careers and costs. (Early on the Smithsonian estimated that an inventory of its human remains would cost more than $1 million.) Supporters argued that money aside, the law was right. Optimists felt certain that NAGPRA invited a new collaborative spirit into conversations among archaeologists, curators, government officials, and native activists, and that their conversations would hatch new ideas that would benefit everyone. In many cases, those predictions have come true. Scholars agreed to return historic remains and to consult with tribal leaders on excavations. Cynics have argued that the law expresses anti-scientific sentiments inspired by shortsighted, politically correct activists, who threaten to destroy valuable data before it has been well studied.[21] (Hrdlička might have argued that the law unnecessarily favors the contemporary remains mixed into burial trenches.) Caricatures of these positions pit hard-hearted scientists against spiritual natives. At the extremes, the caricatures contain a germ of truth. But only a germ, since neither science nor spirit is so simple.

History suggests a subtler story that sets NAGPRA and the collection and display of the native dead as a chapter in the country's struggle for equality. Some see in the collected dead signs of the unhappy triumph of an unethical secular science, a world stripped of its enchanting mysteries and deeper meanings. And comments in the correspondence of collectors, who congratulated themselves on their brave new science, support this idea. Sometimes their bravery seems absurd. Think of Morton ready to take on the biblical account of creation and imagining himself as Galileo's intellectual heir. Think of Gliddon and Morton laughing at a superstitious Egyptian servant. Or Agassiz asking for "handsome fellows." Or every grave-robbing collector who boasted that he had outfoxed grieving friends and relatives who tried to protect a corpse.

But something more than secularism seems to be at work in each of these cases. It wasn't secular science that blinded tearful Captain Wilkes to differences between his nephew's body and the body of Veidovi, but rather a world where he had more power than a Fijian, where he could act on his grief and his spiritual needs, and a Fijian, made his captive, could not. Nor is it an empty secular worldview that makes the Surgeon General's 1868 request for Native American crania seem so mean. What is striking is the contrast between that order's effects on native communities (where the living watched as the bodies of ancestors and kin were dug up and packed off to Washington), and Washington's simultaneous efforts to sort and rebury the Civil War's dead, a process a grieving country believed would help heal war wounds still raw in Northern communities. Congress allocated resources for reburial, expressing a political will to create sacred grounds that would knit the war's dead into the country's future.

The native dead have earned a place in that future too. But it has been long and slow in coming, interrupted by strange sojourns in places like Morton's shelves, the Army Medical Museum's cabinets and labs, and Hrdlička's drawers. In the 1830s and 1840s, collectors descended on burial grounds as the government pushed Indians farther and farther west, sure that Native Americans had no place in America's future. Postwar collectors continued that work on the western plains, in Utah's mountains, and on the California coast. Removing the dead erased markers of past settlement and helped open the land for American farmers and town builders. Sad Professor McGill knew this future threatened to forget his son too.

Yet in one of the many ironies that run through America's history, collectors and catalogers have preserved traces of both young Dr. McGill and clues to the stories of thousands and thousands of unburied and collected dead. Collectors robbed graves, insulted survivors, turned humans into numbers and their remains into specimens, curiosities, trophies, and pen racks. They also stored their memories, recorded their customs, and left signs of remarkable stories for others to follow. The skulls have become embattled objects, but in the long run it might be easier to rebury the collected skulls than to rid our culture of the racist ideas that some extracted from collected skulls. Or in a better world, there should be ways for the two to happen together.

Notes

INTRODUCTION

1. Samuel George Morton, *Crania Americana; or, A comparative view of the skulls of various aboriginal nations of North and South America* (Philadelphia: J. Dobson, 1839), 5. Scholarship on Morton's work includes William Stanton, *The Leopard's Spots: Scientific Attitudes toward Race in America* (Chicago: University of Chicago Press, 1972); Bruce Dain, *A Hideous Monster of the Mind: American Race Theory in the Early Republic* (Cambridge, MA: Harvard University Press, 2003); and Winthrop P. Jordan, *White over Black: American Attitudes toward the Negro, 1550–1812* (Chapel Hill: University of North Carolina Press, 1968). All cover the intellectual history of "scientific racism." See also George Fredrickson, *The Black Image in the White Mind: The Debate on Afro-American Character and Destiny, 1817–1914* (Middletown, CT: Wesleyan University Press, 1971); Reginald Horsman, *Race and Manifest Destiny: The Origins of American Racial Anglo-Saxonism* (Cambridge, MA: Harvard University Press, 1981); and Mia Bay, *The White Image in the Black Mind: African American Ideas about White People, 1830–1925* (New York: Oxford University Press, 2000).

2. Henry S. Patterson, M.D., "Memoir of the Life and Scientific Labors of Samuel George Morton," in *Types of Mankind; or, Ethnological Researches, Based upon the Ancient Monuments, Painting, Sculptures, and Crania of the Races and upon Their Natural, Geographical, Philological and Biblical History*, ed. Josiah C. Nott and George R. Gliddon (Philadelphia: Lippincott, Grambo, & Co., 1854), xxxviii. The story of Morton and his skulls belongs in the chronicles of a "disenchanted world," the history of a particular style of rationality that has characterized Western science. That story's nineteenth-century American chapter interests me. Readers wanting the bigger picture can begin with Michel Foucault, *The Order of*

Things: An Archaeology of the Human Sciences (New York: Vintage, 1973 [1966]); Paula Findlan, *Possessing Nature: Museums, Collecting, and Scientific Culture in Early Modern Italy* (Berkeley: University of California Press, 1994); and Steven Shapin, *A Social History of Truth: Civility and Science in Seventeenth-Century England* (Chicago: University of Chicago Press, 1994). On collecting, see Jean Baudrillard, "The System of Collecting," trans. Roger Cardinal, in *The Cultures of Collecting*, ed. John Elsner and Roger Cardinal (Cambridge, MA: Harvard University Press, 1994), 8; and John V. Pickstone, *Ways of Knowing: A New History of Science, Technology, and Medicine* (Chicago: University of Chicago Press, 2000), 60–82; 106–19.

3. The discussion of biological differences inspired paleontologist Stephen Jay Gould's *The Mismeasure of Man* (New York: Norton, 1996), perhaps the most careful study of Morton's methods.

4. David Hurst Thomas, *Skull Wars: Kennewick Man, Archaeology, and the Battle for Native Identity* (New York: Basic Books, 2000); Robert Bieder, *Science Encounters the Indian, 1820–1880: The Early Years of American Ethnology* (Norman: University of Oklahoma Press, 1986). Orin Starn, *Ishi's Brain: In Search of America's Last "Wild" Indian* (New York: Norton, 2004) and Kenn Harper, *Give Me My Father's Body: The Life of Minik, the New York Eskimo* (South Royalton, VT: Steerforth Press, 2000) describe two recent repatriation cases.

5. Alexis de Tocqueville saw the inequality in the treatment of black bodies. "When a Negro dies, his bones are cast aside, and the distinctions of condition prevail even in the equality of death. Thus a Negro is free, but he can share neither the rights, nor the pleasures, nor the labor, nor the afflictions, nor the tomb of him whose equal he has been declared to be; and he cannot meet him upon fair terms in life or in death." *Democracy in America* (New York: Bantam Dell, 2000 [1835]), 417.

6. On the use of bodies—dead and alive—to construct scientific and cultural authority, see Londa Schiebinger, *Nature's Body: Gender in the Making of Modern Science* (Boston: Beacon Press, 1993). On the development of anatomy, see Michael Sappol, *A Traffic of Dead Bodies: Anatomy and Embodied Social Identity in Nineteenth-Century America* (Princeton, NJ: Princeton University Press, 2002); and Ruth Richardson, *Death, Dissection, and the Destitute* (Chicago: University of Chicago Press, 1987).

7. On slavery and burial practices, see Stephanie Smallwood, *Saltwater Slavery: A Middle Passage from Africa to American Diaspora* (Cambridge, MA: Harvard University Press, 2007), 140–41; and Vincent Brown, *The Reaper's Garden: Death and Power in the World of Atlantic Slavery* (Cambridge, MA: Harvard University Press, 2008), especially 5–6, 157–200. I also follow Claudio Lomnitz, *Death and the Idea of Mexico* (New York: Zone Books, 2005); Joseph Roach, *Cities of the Dead: Circum-Atlantic Performance* (New York: Columbia University Press, 1996); and Robert Pogue Harrison, *The Dominion of the Dead* (Chicago: University of Chicago Press, 2005), on funerals and burial practices.

8. Drew Gilpin Faust, *This Republic of Suffering: Death and the American Civil War* (New York: Knopf, 2008).

9. In the 1980s, the Native American Rights Fund estimated that there were as many as 600,000 pieces of human remains (including thousands of skulls) in American collections—in libraries, museums, historical societies, universities, anatomical collections, and private cabinets. See Douglas Preston, "Skeletons in Our Museums' Closets," *Harper's Monthly*, February 1989, 67. Walter Echo-Hawk edited a special edition of *American Indian Culture and Research Journal* 16, no. 2 (1992), which explores issues behind the Native American Graves Protection and Repatriation Act. See also Kathleen S. Fine-Dare, *Grave Injustice: The American Indian Repatriation Movement and NAGPRA* (Lincoln: University of Nebraska Press, 2002); and Roger C. Echo-Hawk and Walter R. Echo-Hawk, *Battlefields and Burial Grounds* (Minneapolis: Lerner Publications, 1994).

10. Recent books that take a humorous turn on unburied bodies (or body parts) include Michael Paterniti, *Driving Mr. Albert: A Trip across America with Einstein's Brain* (New York: Dial, 2000); Brian Burrell, *Postcards from the Brain Museum: The Improbable Search for Meaning in the Matter of Famous Minds* (New York: Broadway Books, 2004); Mark Svenvold, *Elmer McCurdy: The Misadventures in Life and Afterlife of an American Outlaw* (New York: Basic Books, 2002); and Paul Collins, *The Trouble with Tom: The Strange Afterlife and Times of Thomas Paine* (New York: Bloomsbury, 2005). See also Mary Roach, *Stiff: The Curious Lives of Human Cadavers* (New York: Norton, 2003); and Heather Pringle, *The Mummy Congress: Science, Obsession, and the Everlasting Dead* (New York: Hyperion, 2001).

CHAPTER ONE

1. William R. Grant, *Sketch of the Life and Character of Samuel George Morton, M.D.: Lecture Introductory to a Course on Anatomy and Physiology in the Medical Department of Pennsylvania College, Delivered October 13, 1851* (Philadelphia, 1852), 4, 11; Patterson, "Memoir," xviii–lvii.

2. George H. Daniels, "The Process of Professionalization in American Science: The Emergent Period, 1820–1860," *Isis* 58, no. 2 (Summer 1967): 150–66; Bruno Latour, *Science in Action: How to Follow Scientists and Engineers through Society* (Cambridge, MA: Harvard University Press, 1987); Laura Dassow Walls, "Textbooks and Texts from the Brooks: Inventing Scientific Authority in America," *American Quarterly* 49, no. 1 (March 1997): 1–2.

3. "Death of Samuel George Morton, M.D.," *Medical Examiner and Record of Medical Science*, June 1851, 384–85. Morton's private autopsy honored the man. It did not carry the insult of public dissections often performed on executed criminals and occasionally on dead celebrities. Sappol, *A Traffic of Dead Bodies*, 103; Benjamin Reiss, "P. T. Barnum, Joice Heth, and Antebellum Spectacles of Race," *American Quarterly* 51, no. 1 (March 1999): 78–107; Benjamin Reiss, *The Showman and the Slave: Race, Death, and Memory in Barnum's America* (Cambridge, MA: Harvard University Press, 2001).

4. "St. Tertullian," *The Recreative Magazine; or, Eccentricities of Literature and Life*, April 15, 1822, 373–89 [American Periodical Series online]. A few years after Morton's death, a Boston paper reminded readers of the sale of Descartes' skull. "Antiquities and Curiosities," *Boston Investigator*, December 19, 1858. Russell

Shorto, *Descartes' Bones: A Skeletal History of the Conflict between Faith and Reason* (New York: Doubleday, 2008), tells the full story of the philosopher's remains.

5. Simon J. Harrison, "Skulls and Scientific Collecting in the Victorian Military: Keeping the Enemy Dead in British Frontier Warfare," *Comparative Studies in Society and History* 50, no. 1 (2008): 298–300. See also Laura Franey, "Ethnographic Collecting and Travel: Blurring Boundaries, Forming a Discipline," *Victorian Literature and Culture* 29, no. 1 (2001): 219–39; Tom Griffiths, *Hunters and Collectors: The Antiquarian Imagination in Australia* (Cambridge: Cambridge University Press, 1996), 28–54; James Axtell and William C. Sturtevant, "The Unkindest Cut; or, Who Invented Scalping," *William and Mary Quarterly* 3rd ser., 37 (1980): 451–72; and Zine Magubane, "Simians, Savages, Skulls, and Sex: Science and Colonial Militarism in Nineteenth-Century South Africa," in *Race, Nature, and the Politics of Difference*, ed. Donald S. Moore, Jake Kosek, and Anand Pandian (Durham, NC: Duke University Press, 2003), 99–121. On the longer history of bodies as sacred objects, see Patrick Geary, "Sacred Commodities: The Circulation of Medieval Relics," in *The Social Life of Things*, ed. Arjun Appadurai (Cambridge: Cambridge University Press, 1986), 169; and Peter Brown, *The Cult of the Saints: Its Rise and Function in Latin Christianity* (Chicago: University of Chicago Press, 1981).

6. Dell Upton, *Another City: Urban Life and Urban Spaces in the New American Republic* (New Haven, CT: Yale University Press, 2008), 203–41; Marilyn Yalom, *The American Resting Place: Four Hundred Years of History through Our Cemeteries and Burial Grounds* (Boston: Houghton Mifflin, 2008), 45–47, 101–5.

7. Laurel Hill historian Carol Yaster gives the story an ironic twist. "There used to be a bust of Morton under the 'canopy,'" she writes, "but all that's left now is the 'chest' portion. Vandals are attracted to heads on monuments." Who knows who has collected Morton's head? (E-mail to author, April 17, 2009.) Colleen McDannell, "The Religious Symbolism of Laurel Hill Cemetery," *Pennsylvania Magazine of History and Biography* 111, no. 3 (July 1987): 275–303; "Laurel Hill Cemetery," *Mechanics' Magazine and Journal of Mechanic's Institute*, January 1837, 9; Nathan Dunn, "Entrance to Laurel Hill Cemetery," *Ladies' Garland and Family Wreath Embracing Tales, Sketches, Incidents, History, Poetry, Music, Etc.*, January 1838, 196. (Thanks to Mel Stein for braving floods and visiting the Morton graves and to Carol Yaster for taking more pictures.) On Morton's funeral, see Grant, *Sketch of the Life*, 5.

8. J. Aitken Meigs, *Catalogue of Human Crania, in the Collection of the Academy of Natural Sciences of Philadelphia* (Philadelphia: Merrihew & Thompson, 1857), 1.

9. On Morton's contribution to the American school of ethnology, see Stanton, *The Leopard's Spots*; Gould, *The Mismeasure of Man*, 82–104; and Jordan, *White over Black*, 216–65. See also Daniel Moore Fisk, "Samuel George Morton," *Dictionary of American Biography*, ed. Dumas Malone (New York: Charles Scribner's Sons, 1934), 7:265–66; and Patterson, "Memoir," xxxi–xxxiii. Latour, *Science in Action*, 215–57, helps picture Morton's Philadelphia as a "centre of calculation."

10. Susan Scott Parrish, *American Curiosity: Cultures of Natural History in the Colonial British Atlantic World* (Chapel Hill: University of North Carolina Press, 2006), 8–9; Thomas P. Slaughter, *The Natures of John and William Bartram* (New

York: Vintage Books, 1996); Dorinda Outram, "New Spaces in Natural History," in *Cultures of Natural History*, ed. N. Jardine, J. A. Secord, and E. C. Spary (Cambridge: Cambridge University Press, 1996), 249–65. Paul Turnbull describes the life of Alexander Berry, an Australian skull collector whose education and practice resembled Morton's. "'Rare Work amongst the Professors': The Capture of Indigenous Skulls within Phrenological Knowledge in Early Colonial Australia," in *Body Trade: Captivity, Cannibalism and Colonialism in the Pacific*, ed. Barbara Creed and Jeanette Hoorn (New York: Routledge, 2001), 7–14.

11. Morton, *Crania Americana*, 186–87. See also Samuel George Morton, "Observations on the size of the Brain in various Races and Families of Man," *Proceedings of the American Philosophical Society*, October 1849, 224.

12. "Query VI: Productions, Mineral, Vegetable, and Animal," in Thomas Jefferson, *Notes on the State of Virginia*, ed. David Waldstreicher (New York: Bedford Books, 2002 [1787]), 117–20.

13. "Query XIV: Laws," in Jefferson, *Notes*, 180.

14. Sappol, *A Traffic of Dead Bodies*, 14–15, 17.

15. Charles D. Meigs, M.D., "Extracts from a Memoir of Samuel George Morton, M.D., Late President of the Academy of Natural Sciences of Philadelphia," *American Journal of Science and Arts*, March 1852, 154; Patterson, "Memoir," xxi. Tables added the power of numbers to Morton's work. For mid-century America's skills in arithmetic, see Patricia Cline Cohen, *A Calculating People: The Spread of Numeracy in Early America* (New York: Routledge, 1999 [1982]).

16. George H. Daniels, *American Science in the Age of Jackson* (New York: Columbia University Press, 1968), 210–11; Simon Baatz, "Philadelphia Patronage: The Institutional Structure of Natural History in the New Republic, 1800–1833," *Journal of the Early Republic* 8 (Summer 1988): 111–38; Patterson, "Memoir," xxiii; Richard Harlan, *Fauna Americana, being a description of the mammiferous animals inhabiting North America* (Philadelphia: A. Finley, 1825); Jean-Nicolas Gannal, *History of embalming, and of preparations in anatomy, pathology, and natural history; including an account of a new process for embalming*, trans. Richard Harlan (Philadelphia: J. Dobson, 1840), 257.

17. Richard Harlan, *Medical and Physical Researches; or, Original memoirs in medicine, surgery, physiology, geology, zoology, and comparative anatomy* (Philadelphia: Bailey, 1835), 24. The exchange with Bachman is in Lester D. Stephens, *Science, Race, and Religion in the American South: John Bachman and the Charleston Circle of Naturalists, 1815–1895* (Chapel Hill: University of North Carolina Press, 2000), 25–26. See also Patsy A. Gerstner, "The Academy of Natural Sciences of Philadelphia, 1812–1850," in *The Pursuit of Knowledge in the Early Republic: American Scientific and Learned Societies from Colonial Times to the Civil War*, ed. Alesandra Oleson and Sanborn C. Brown (Baltimore: Johns Hopkins University Press, 1976), 186–87; and Brett Mizelle, "'Man Cannot Behold It without Contemplating Himself': Monkeys, Apes, and Human Identity in the Early American Republic," *Pennsylvania History* 66 (1999): 144–73.

18. "The Fire at Wetherill's White Lead Manufactory," *North American*, June 18, 1839, col. B. 19th Century U.S. Newspapers, accessed July 8, 2008.

19. Daniels, *American Science*, 210–11; J. A. Meigs, *Catalogue*, 27, 47, 55.

20. Morton's name does not appear on the list of subscribers to a compilation of Harlan's papers published in 1835. Debates about the origins of racial differences could have played a part in their quarrel. While Morton moved toward polygeny, Harlan remained a monogenist, certain that all men are "*varieties of one original stock*." Harlan, *Medical and Physical Researches*, xxiv. The list of subscribers appears on 522.

21. Patterson, "Memoir," xviii. See Dana D. Nelson, *National Manhood: Capitalist Citizenship and the Imagined Fraternity of White Men* (Durham, NC: Duke University Press, 1998), 102–34.

22. Robert Mudie, *The Modern Athens: A Dissection and Demonstration of Men and Things in the Scotch Capital* (London: Knight and Lacey, 1825), 210; Sappol, *A Traffic of Dead Bodies*, 50–53.

23. Mudie, *The Modern Athens*, 150, 158; Lisa Rosner, *Medical Education in the Age of Improvement: Edinburgh Students and Apprentices, 1760–1820* (Edinburgh: Edinburgh University Press, 1991), 25; C. Meigs, "Extracts," 157.

24. James Rush quoted in Rosner, *Medical Education*, 39. On the origin of students, see ibid., 17; on student life in Edinburgh, see Adrian Desmond and James Moore, *Darwin: The Life of a Tormented Evolutionist* (New York: Norton, 1991), 25.

25. C. Meigs, "Extracts," 159, quotes Morton's diary from his student years. Rosner, *Medical Education*, 3–4, 24, 46.

26. Anand C. Chitnis, "Medical Education in Edinburgh, 1790–1826, and Some Victorian Social Consequences," *Medical History*, 17, no. 2 (April 1973): 174–75, 179. Barclay's popularity put pressure on Edinburgh's cadaver supply. There were too few executions to insure a steady stream of criminal dead, the only bodies legally available for dissection, and Barclay turned to "resurrectionists" (grave robbers who supplied corpses to medical schools) and had bodies shipped up from London and Liverpool. See John Struthers, *Historical Sketch of the Edinburgh Anatomical School* (Edinburgh: Maclachlan and Stewart, 1867), 59–70. On Edinburgh and the social history of anatomy more generally, see Richardson, *Death, Dissection, and the Destitute*, 3–52; and Sappol, *A Traffic of Dead Bodies*, 49–51.

27. C. Meigs, "Extracts," 158; Anand C. Chitnis, "The University of Edinburgh's Natural History Museum and Huttonian-Wernerian Debate," *Annals of Science* 26, no. 2 (June 1970): 85–94; Carla Yanni, *Nature's Museums: Victorian Science and the Architecture of Display* (Baltimore: Johns Hopkins University Press, 1999), 91–109; Alexander Grant, *The Story of the University of Edinburgh during Its First Three Hundred Years* (London: Longmans, Green and Co., 1884), 376.

28. Patterson, "Memoir," xxiv–xxv.

29. Desmond and Moore, *Darwin*, 21–30.

30. Maria Audubon, *Audubon and His Journals* (Gloucester, MA: Peter Smith, 1972), 160; J. J. Audubon to Samuel George Morton, June 25, 1838, Samuel George Morton Papers, American Philosophical Society (hereafter Morton Papers, APS); J. A. Meigs, *Catalogue*, 90, 100–101.

31. M. Audubon, *Audubon and His Journals*, 191–92.

32. Patterson, "Memoir," xxxii.

33. Ibid., xxxii. E. P. Thompson, *The Making of the English Working Class* (London: V. Gollancz, 1963), sparked scholarly interest in phrenology as part of the intellectual repertoire of working-class Britons. See C. N. Cantor, "Phrenology in Early Nineteenth-Century Edinburgh: An Historiographical Discussion," *Annals of Science* 32 (1975): 195–218; Steven Shapin, "Phrenological Knowledge and Social Structure of Early Nineteenth-Century Edinburgh," *Annals of Science* 32 (1975): 219–43; Terry Parssinen, "Popular Science and Society: The Phrenology Movement in Early Victorian Britain," *Journal of Social History* 8 (Fall 1974): 1–20; Angus McLaren, "Phrenology: Medium and Message," *Journal of Modern History* 46 (March 1974): 86–97; and Roger Cooter, *The Cultural Meaning of Popular Science: Phrenology and the Organization of Consent in Nineteenth-Century Britain* (Cambridge: Cambridge University Press, 1984). On Combe, see David Stack, *Queen Victoria's Skull* (London: Hambeldon Continuum, 2008). Charles Colbert, *A Measure of Perfection: Phrenology and the Fine Arts in America* (Chapel Hill: University of North Carolina Press, 1997), is an excellent account of phrenology in the United States, but see also John Davis, *Phrenology, Fad and Science: A 19th-Century American Crusade* (New Haven, CT: Yale University Press, 1955); and Madeleine B. Stern, *Heads and Headlines: The Phrenological Fowlers* (Norman: University of Oklahoma Press, 1971).

34. C. Meigs, "Extracts," 160; "Death of Dr. Morton," *American and Gazette*, May 16, 1851. On Philadelphia medicine, see Richard H. Shyrock, "The Advent of Modern Medicine in Philadelphia, 1800–1850," *Yale Journal of Biology and Medicine* 13 (July 1941): 725; and Sappol, *A Traffic of Dead Bodies*, 3–4, 64–65.

35. Samuel George Morton, "History of the Academy of Natural Sciences of Philadelphia," *American Quarterly Review* (May 1841): 433–38. See Paul Semonin, "'Nature's Nation': Natural History as Nationalism in the New Republic," *Northwest Review* 30 (1992): 6–41; Outram, "New Spaces in Natural History," 249–65; and Lorraine Daston, "The Factual Sensibility," *Isis* 79 (1988): 452–67.

36. W. S. W. Ruschenberger, *A Notice of the Origins, Progress, and Present Condition of the Academy of Natural Sciences of Philadelphia* (Philadelphia: Collins, Printer, 1860), 68; Robert V. Bruce, *The Launching of Modern American Science, 1846–1876* (New York: Knopf, 1987), 36; Baatz, "Philadelphia Patronage," 24–26; Gerstner, "The Academy of Natural Sciences," 174–93.

37. Slaughter, *The Natures of John and William Bartram*, xv, 224–57; Charles Coleman Sellers, *Mr. Peale's Museum: Charles Willson Peale and the First Popular Museum of Natural Science and Art* (New York: Norton, 1980); Toby A. Appel, "Science, Popular Culture and Profit: Peale's Philadelphia Museum," *Journal of the Society for the Bibliography of Natural History* 9, no. 4 (April 1980): 619–34; Steven Conn, *Museums and American Intellectual Life, 1876–1926* (Chicago: University of Chicago Press, 1998), 32–73; Laura Rigal, *The American Manufactory: Art, Labor, and the World of Things in the Early Republic* (Princeton, NJ: Princeton University Press, 1998); Paul Semonin, *American Monster: How the Nation's First Prehistoric Creature Became a Symbol of National Identity* (New York: New York University Press, 2000).

38. George Ord quoted in Baatz, "Philadelphia Patronage," 28.

39. Patterson, "Memoir," xxvi; Baatz, "Philadelphia Patronage," 34.

40. On the "sedentary naturalist," see Outram, "New Spaces in Natural History," 249–65; and Walls, "Textbooks and Texts from the Brooks."

41. Samuel G. Morton, "Diary, 1833–ca. 1837," n.p., Morton Papers, APS.

42. Morton, *Crania Americana*, 3. On origin stories, see Nancy Shoemaker, "How Indians Got to Be Red," *American Historical Review* 102, no. 3 (June 1997): 636.

43. Morton's opinion is reprinted in C. Meigs, "Extracts," 172. See also Samuel George Morton, *Brief remarks on the diversities of the human species: and on some kindred subjects: being a introductory lecture delivered before the class of Pennsylvania Medical College, in Philadelphia, November 1, 1842* (Philadelphia: Merrihew & Thompson, Printers, 1842); and Samuel George Morton, *Additional observations on hybridity in animals, and on some collateral subjects: being a reply to the objections of the Rev. John Bachman* (Charleston: Walker & James, 1850).

44. Moravian missionary John Heckewelder described Nanticoke people passing through Bethlehem, Pennsylvania, "loaded with the bones of their dead friends, some of which were in so recent a state as to taint the air as they passed." Morton, *Crania Americana*, 81.

45. C. Meigs, "Extracts," 162. See also Schiebinger, *Nature's Body*, 115–34.

46. Petrus Camper, *The works of the late Professor Camper: On the connexion between the science of anatomy and the arts of drawing, painting, statuary, &c., &c.*, ed. T. Cogan (London, 1821), 13. The quote from Cuvier appears in Georges Cuvier, "Note instructive sur les recherches à faire relativement aux différences anatomiques des diverses races d'hommes," in G. Hervé, "A la recherche d'un manuscrit: Les instructions anthropologique de G. Cuvier pour le Voyage du 'Géographe' et du 'Naturaliste' aux Terres Australes," *Revue de l'École d'Anthropologie de Paris* 20 (1910): 304.

47. Camper, *Works*, 9, 38, 50; James Cowles Prichard, *Researches into the Physical History of Man* (London: John and Arthur Arch, 1813), 48.

48. Nancy Stepan, *The Idea of Race in Science: Great Britain, 1800–1960* (Hamden, CT: Archon Books, 1982), 28; James Aitken Meigs, *Hints to craniographers, upon the importance and feasibility of establishing some uniform system by which the collection and promulgation of craniological statistic and the exchange of duplicate crania may be promoted* (Philadelphia: Merrihew & Thompson, 1858), 4.

49. Camper, *Works*, 8–9; George W. Stocking Jr., *Race, Culture, and Evolution: Essays in the History of Anthropology* (New York: Free Press, 1968), 29.

50. Johann Friedrich Blumenbach, *The Anthropological Treatises of J. Friedrich Blumenbach*, trans. Thomas Bendyshe (London: Longman, Green, Roberts, & Green, 1865), 55–56, 276, 299; David Bindman, *Ape to Apollo: Aesthetics and the Idea of Race in the 18th Century* (Ithaca, NY: Cornell University Press, 2002), 202–6; Schiebinger, *Nature's Body*, 152–54. Gould, *The Mismeasure of Man*, 401–12, describes the consequences of Blumenbach's "geometry" of five races.

51. Cuvier, "Note," 298, 305–6; Stocking, *Race, Culture, and Evolution*, 30.

52. Patterson, "Memoir," xxxiii.

53. Ibid., xviii; Samuel George Morton, "Introduction," in J. A. Meigs, *Catalogue*, 13.

54. Samuel George Morton to T. A. Conrad, December 17, 1832, Morton Letter Book, Manuscripts Division, Department of Rare Books and Special Collections, Princeton University Library (hereafter cited as Morton Letter Book).

55. Samuel George Morton to Dr. Ezekial Skinner, December 19, 1835, Morton Letter Book.

56. J. A. Meigs, *Catalogue*, 94. S. M. E. Goheen also donated African reptiles to American herpetology collections. On violence in Liberia, see Claude A. Clegg III, *The Price of Liberty: African Americans and the Making of Liberia* (Chapel Hill: University of North Carolina Press, 2004), 247–28. Settler Sion Harris describes taking "Goterah's head" in a letter to Samuel Wilkeson in *Slaves No More: Letters from Liberia, 1833–1869*, ed. Bell I. Wiley (Lexington: University Press of Kentucky, 1980), 220–23. Morton's collecting could play a part in struggles over African witchcraft. Dr. Robert McDowell sent Morton the skull of a man from the Bassa tribe. "The skull of an African *Gree-gree man*, or doctor. For committing some crime he was tried by the ordeal of drinking *red-wood water*, and found guilty, was cut in pieces, and thrown into the St. John's river, Grand Bassa, where his skull was found—a very good specimen of the Bassa tribe." J. A. Meigs, *Catalogue*, 92.

57. B. F. Alter to Samuel George Morton, July 22, 1845, Samuel George Morton Papers, Library Company of Philadelphia (hereafter cited as Morton Papers, LCP).

58. Ruschenberger, *Notice*, 46. Morton may have been looking to Lord Kames, who had posited multiple acts of creation to solve the puzzle of human difference. See Paul B. Wood, "The Science of Man," in *Cultures of Natural History*, ed. Jardine, Secord, and Spary, 204.

59. J. A. Meigs, *Catalogue*, 45; Marmaduke Burrough to Samuel George Morton, March 15, 1836, Morton Papers, APS. Burrough (1814–1843) served as United States counsel at Calcutta and Vera Cruz. Like many of Morton's friends and correspondents, Burrough moved along a backwater circuit of history. He toured the United States with a rhinoceros and an orangutan. A phrenologist read his head in December 1841. "After exploring his bumps, the professor pronounced him to be inclined to skepticism in religion, to be fond of travelling and of pets, to have a good memory of the face of any country he might have passed, but a poor one for circumstances, to have great concentrativeness, and strong love of order (which might be inferred from his dress) to be fond of ladies etc. The subject acknowledged the description to be true so far as he knew himself." Isaac Mickle, *A Gentleman of Much Promise: The Diary of Isaac Mickle, 1837–1845*, ed. Philip English Mackey (Philadelphia: University of Pennsylvania Press, 1977), 1:243.

60. J. A. Meigs, *Catalogue*, 59.

61. Marmaduke Burrough to Samuel George Morton, March 13, 1835, Morton Papers, APS; J. A. Meigs, *Catalogue*, 25. For more on Pierce (or Pearce, as he appears in some accounts), see Helen MacDonald, *Human Remains: Dissection and Its Histories* (New Haven, CT: Yale University Press, 2005), 101–2. Paul Turnbull reports that colonial surgeon James Scott dissected Pierce's body at Hobart, part of his official punishment. After Scott's dissection, the skull began a last long

journey to Philadelphia. Turnbull, "'Rare Work,'" 12. The case for repatriation of Native American remains developed out of the history of disputes between Native Americans and archaeologists and collectors. Other groups whose skulls were collected in the United States have not pushed for repatriation.

62. John L. Stephens, *Incidents of Travel in Yucatan*, 2 vols. (New York: Dover, 1963 [1843]), 1:143, 175, 200. Stephens brought back one skeleton so shattered he wrapped its pieces in a pocket handkerchief. A less experienced man would have tossed it out, but Morton got out his glue pot and went to work, reassembling the fragments into the skeleton of a young Indian woman, whose teeth were white and fresh. The account appears in Patterson, "Memoir," xxxix. See also Samuel George Morton, *An Inquiry into the Distinctive Characteristics of the Aboriginal Race of America* (Philadelphia: John Pennington, 1844), 39.

63. Samuel George Morton to Don José María Vargas, January 1, 1836, Morton Letter Book. José María Vargas to Samuel George Morton, April 2, 1838, Morton Papers, APS; J. A. Meigs, *Catalogue*, 75, 101; Morton, *Crania Americana*, 237.

64. J. A. Meigs, *Catalogue*, 73–74. On gender and specimens, see Elizabeth Fee, "19th-Century Craniology: The Study of the Female Skull," *Bulletin of the History of Medicine* 53 (1979): 415–33.

65. C. Meigs, "Extracts," 164.

66. Doornik traveled around the Indonesian archipelago, trading for skulls while companions traded for coffee and spices. He is listed among the passengers on the brig *Maria* in 1814 and 1815. "Ships Arrivals and Departures 1814–1815," The Java Half-Yearly Almanac and Directory for 1815, http://home.hccnet.nl/wer.davies/jbrifoad.html (accessed January 14, 2009). He left the Dutch East Indies in 1826 under fire for an essay criticizing the Dutch governor general. Doornik arrived in the United States in 1827. See H. A. M. Snelders, "Doornik, Jacob Elisa," in *History of Physical Anthropology: An Encyclopedia*, ed. Frank Spencer (New York: Garland, 1997), 1:352–53. On human remains in Java, see Stubbins Ffirth, M.D., "Some Accounts of Batavia and Its Sources of Unhealthfulness," *Philadelphia Medical Museum* 2 (1806): 48.

67. Donald M. Scott, "The Popular Lecture and the Creation of a Public in Mid-Nineteenth-Century America," *Journal of American History* 66, no. 4 (March 1980): 719–809. Charles Pickering to Morton, December 9, 1842, Samuel George Morton Microfilm, APS; Baatz, "Philadelphia Patronage," 20.

68. J. E. Doornik to Samuel George Morton, June 23, 1835, Morton Papers, APS.

69. Doornik to Morton, June 19, June 23, June 28, July 3, July 11, July 21, 1835, Morton Papers, APS. Morton to Doornik, July 10, 1835, Morton Letter Book. For skulls from Doornik, see J. A. Meigs, *Catalogue*, 23, 54, 97, 102. Benjamin H. French obtained several of Doornik's skulls from Dr. Jones of New Orleans and sent them to Morton. Benj. H. French to Morton, September 29, 1842, and October 18, 1845, Morton Papers, LCP.

70. C. Meigs, "Extracts," 153–78; "Eulogy on the Late Dr. Morton," *New York Times*, January 17, 1852, 6.

71. Sanford B. Hunt, "Samuel George Morton," in *Lives of Eminent American Physicians and Surgeons of the Nineteenth Century*, ed. Samuel D. Gross (Phila-

delphia: Lindsay & Blakiston, 1861), 591; W. S. W. Ruschenberger, *The Claims of the Academy of Natural Sciences of Philadelphia to Public Favor* (Philadelphia: Collins, Printer, 1871), 9, 12. On Agassiz, see J. A. Meigs, *Catalogue*, 3–4; and Patterson, "Memoir," xxx.

72. J. A. Meigs, *Catalogue*, 52; David Chapin, *Exploring Other Worlds: Margaret Fox, Elisha Kent Kane, and the Antebellum Culture of Curiosity* (Amherst: University of Massachusetts Press, 2004), 27–28.

73. Hunt, "Samuel George Morton," 591, 596.

74. Carl Berger, "Sir Daniel Wilson," *Dictionary of Canadian Biography* (Toronto: University of Toronto Press, 1891–1900), 12:1110.

75. W. J. Hoffman, M.D., *Report on the Chaco Cranium* (Washington, DC: Government Printing Office, 1878), 457. On skull borrowing, see Academy of Natural Sciences meeting minutes, January 8, February 12, March 5, and March 14, 1889, Archives of the Academy of Natural Sciences, Philadelphia, PA.

76. Daniel Brinton, *Report upon the Collections Exhibited at the Columbian Historical Exhibition at Madrid* (Washington, DC: Government Printing Office, 1895), 23, 194–201, 203–5. On March 14, 1893, members of the academy also authorized a loan of "certain crania" to Professor Putnam for exhibition at Chicago, distinguishing between the Columbian Exposition at Chicago, which illustrated progress on the continent, and the exhibition at Madrid, which illustrated "primitive American life and the history of the period of discovery and conquest." Barr Ferree, "The Historical American Exhibition at Madrid," *Nation* 53 (December 24, 1891): 487. See also "D.D.," "The United States at the Historico-American Exposition," *Nation* 55 (November 10, 1892): 349–50. On the Spanish politics and the exhibition, see Michel-Rolph Trouillot, *Silencing the Past: Power and the Production of History* (Boston: Beacon Press, 1995), 119–36. The German anthropologist Rudolf Virchow also imagined native skulls as a tribute to Columbus. He dedicated his *Crania Ethnica Americana* to Columbus's memory––"Zur Erinnerung an Columbus und die Entdeckung Amerikas" (Berlin: A. Asher and Co., 1892). "Medals Awarded at Madrid," *New York Times*, January 19, 1893, 6.

77. The last two decades have witnessed dead bodies moving in Europe, Africa, Australia, New Zealand, as well as the United States. On Eastern Europe, see Katherine Verdery, *The Political Lives of Dead Bodies: Reburial and Postsocialist Change* (New York: Columbia University Press, 1999). On Africa, see Clifton Crais and Pamela Scully, *Sara Baartman and the Hottentot Venus: A Ghost Story and a Biography* (Princeton, NJ: Princeton University Press, 2009).

CHAPTER 2

1. George Combe, *Notes on the United States of North America during a Phrenological Visit in 1838–9–40*, 2 vols. (Philadelphia: Carey & Hart, 1841), 2:175, describes December weather in Philadelphia. Cornelius J. Brosnan, *Jason Lee, Prophet of the New Oregon* (Rutland, VT: Academy Books, 1985 [1932]), 104–41; Morton, *Crania Americana*, 206.

2. The account of the Boston meeting appears in *Zion's Herald*, January 29, 1839, reprinted in Brosnan, *Jason Lee*, 121–22.

3. Thomas L. McKenney and James Hall, *The Indian Tribes of North America with Biographical Sketches and Anecdotes of the Principal Chiefs*, 3 vols. (repr.; Edinburgh: John Grant, 1934), 2:284; Charles D. Meigs, *A Memoir of Samuel Morton, M.D., Late President of the Academy of Natural Sciences of Philadelphia* (Philadelphia: T.K. and P.G. Collins, Printers, 1851), 7. On the place of curiosity in the development of natural history, see Parrish, *American Curiosity*, 57–64; and Lorraine Daston and Katharine Park, *Wonders and the Order of Nature: 1150–1750* (New York: Zone Books, 2001), 311–28.

4. See Daniels, *American Science in the Age of Jackson*, 1–62; D. Graham Burnett, *Trying Leviathan: The Nineteenth-Century Court Case that Put the Whale on Trial and Challenged the Order of Nature* (Princeton, NJ: Princeton University Press, 2007); and Laura Dassow Walls, *Emerson's Life in Science: The Culture of Truth* (Ithaca, NY: Cornell University Press, 2003). On circles, networks, or associations of scientists, see Latour, *Science in Action*, 202, 232–41; and Dana D. Nelson, "'No Cold or Empty Heart': Polygenesis, Scientific Professionalization, and the Unfinished Business of Male Sentimentalism," *differences* 11, no. 3 (Fall 1999): 29–56.

5. Robert J. Loewenberg, *Equality on the Oregon Frontier: Jason Lee and the Methodist Mission* (Seattle: University of Washington Press, 1976).

6. J. W. Redfield, "Among the Flatheads," *American Phrenological Journal* 40, no. 3 (September 1864): 74.

7. Morton to J. N. Nicollet, March 8, 1840, Morton Papers, LCP; "Crania Americana," *Journal of Belles Lettres*, February 27, 1838, 9.

8. Daniel Wilson, "Illustrations of the Significance of Certain Ancient British Skull Forms," *Canadian Journal* 42 (March 1863): 22.

9. See Herman J. Viola, *The Indian Legacy of Charles Bird King* (Washington, DC, and New York: Smithsonian Institution Press and Doubleday, 1976); and Herman J. Viola, *Thomas L. McKenney: Architect of America's Early Indian Policy, 1816–1830* (Chicago: Sage, 1974).

10. Historians have worked to recover the agency of "racialized" performers like Brooks. See Reiss, *The Showman and the Slave*; James W. Cook Jr., "Of Men, Missing Links, and Nondescripts: The Strange Career of P.T. Barnum's 'What Is It?' Exhibition," in *Freakery: Cultural Spectacles of the Extraordinary Body*, ed. Rosemarie Garland-Thomson (New York: New York University Press, 1996), 139–57; L. G. Moses, *Wild West Shows and the Images of American Indians, 1883–1933* (Albuquerque: University of New Mexico Press, 1999); and Philip J. Deloria, *Indians in Unexpected Places* (Lawrence: University Press of Kansas, 2005).

11. The portrait of "Stumanu" does not appear in the "Catalogue of Known Works" in Andrew J. Cosentino, *The Paintings of Charles Bird King (1785–1862)* (Washington, DC: Smithsonian Institution Press, 1977), 121–206. In their 1934 edition of McKenney and Hall, Frederick Webb Hodge and David I. Bushnell Jr. list the artist as "unknown" and note that the portrait is not mentioned in William J. Rhees, *An Account of the Smithsonian Institution, Its Founder, Building, Operations, etc., Prepared from the Reports of Prof. Henry to the Regents, and Other Authentic Sources* (Washington, DC: Thomas M'Gill, printer, 1859), 55–58. F. W. Hodge,

"Introduction," in McKenney and Hall, *The Indian Tribes of North America*, 1:xlvi–liii, 2:287n8. On the blanket, see George T. Emmons, "The Chilkat Blanket, with Notes on the Blanket Design by Franz Boas," *Memoirs of the American Museum of Natural History* 3, no. 4 (1907): 329–401; and Cheryl Samuel, *The Chilkat Dancing Blanket* (Norman: University of Oklahoma Press, 1989). The comment on the blanket appears in "Oregon Mission Concluded," *Boston Recorder* 24 (February 15, 1839): 26.

12. McKenney and Hall, *The Indian Tribes of North America*, 2:277–79.

13. Robert Boyd, *The Coming of the Spirit of Pestilence: Introduced Infectious Diseases and Population Decline among Northwest Coast Indians, 1774–1874* (Vancouver: University of British Columbia Press, 1999), 84–115; Herbert C. Taylor Jr. and Lester L. Hoaglin Jr., "The 'Intermittent Fever' Epidemic of the 1830's on the Lower Columbia River," *Ethnohistory* 9 (Spring 1962): 160–78; Sherburne Cook, "The Epidemic of 1830–1833 in California and Oregon," *University of California Publications in American Archaeology and Ethnography* 43, no. 3 (1955): 303–25; Robert H. Ruby and John A. Brown, *The Chinook Indians: Traders of the Lower Columbia River* (Norman: University of Oklahoma Press, 1976), 185–200.

14. McKenney and Hall, *The Indian Tribes of North America*, 2:284; Cyrus Shepard, letter January 10, 1835, printed in *Zion's Herald*, October 28, 1835, 170, and quoted in Brosnan, *Jason Lee*, 76.

15. McKenney and Hall, *The Indian Tribes of North America*, 2:284; Shepard, quoted in Brosnan, *Jason Lee*, 76; Cyrus Shepard, "Letter from the Oregon Mission," *Zion's Herald* (July 19, 1837): 8, 29. On Lucy Hedding's death, see Gustavus Hines, *A Voyage Round the World: With a history of the Oregon mission* (Buffalo: G. H. Derby, 1850), 17.

16. Historian Gray H. Whaley describes the mission's "grim ledger." Missionaries weighed the costs of feeding and clothing the children against their productive activities. Gray H. Whaley, "'Trophies' for God: Native Mortality, Racial Ideology, and the Methodist Mission of Lower Oregon, 1834–1844," *Oregon Historical Quarterly*, Spring 2006, http://www.historycooperative.org/journals/ohq/107.1/whaley.html (accessed December 1, 2006), para. 32.

17. Claude E. Schaeffer, "William Brooks, Chinook Publicist," *Oregon Historical Quarterly* 64 (1963): 41–54; Robert Moulton Gatke, "The First Indian School of the Pacific Northwest," *Oregon Historical Quarterly* 23 (1922): 70–83; Shepard, "Letter from the Oregon Mission," 29.

18. Jason Lee, "Diary of Rev. Jason Lee," *Quarterly of the Oregon Historical Society* 17 (December 1916): 405, 416; Townsend describes Lee in his *Narrative of a Journey across the Rocky Mountains to the Columbia River and a Visit to the Sandwich Islands, Chili, &c., with a Scientific Appendix* (Corvallis: Oregon State University Press, 1999 [1839]), 12.

19. Loewenberg, *Equality on the Oregon Frontier*, 64; Brosnan, *Jason Lee*, 93, 97; Whaley, "'Trophies' for God," para. 42.

20. Lee, "Diary," 409–10, 413.

21. Ibid., 399; Brosnan, *Jason Lee*, 93.

22. Hines, *Voyage*, 31; Lee, "Diary" 425.

23. Brosnan, *Jason Lee*, 100, 108; George Catlin, "For the Commercial Advertiser," *New-York Spectator*, November 28, 1839; Brian W. Dippie, *Catlin and His Contemporaries: The Politics of Patronage* (Lincoln: University of Nebraska Press, 1990), 47–95.

24. *Christian Advocate and Journal* 13 (February 15, 1839): 102; (March 1, 1839): 109; and 14 (October 4, 1839): 25, quoted in Brosnan, *Jason Lee*, 108, 110, 111, 114, 116; *Zion's Herald*, February 13, 1839, 26, quoted in Brosnan, *Jason Lee*, 118; Hines, *Voyage*, 37.

25. *Christian Advocate and Journal* 14 (October 4, 1839): 25, quoted in Brosnan, *Jason Lee*, 111.

26. Jason Lee, "Some Farther Account of Wm. Brooks, the Flathead Indian," *Western Christian Advocate* 6, no. 26 (October 18, 1839): 104; Brosnan, *Jason Lee*, 104; H. K. Hines, *Missionary history of the Pacific Northwest, containing the wonderful story of Jason Lee, with sketches of many of his co-laborers, all illustrating life on the plains and in the mountains in pioneer days* (Portland: H. K. Hines, 1899), 194.

27. *Oregonian and Indian's Advocate* 1 (January 1839): 125–27, quoted in Brosnan, *Jason Lee*, 121; *Zion's Herald*, February 13, 1839, 27, quoted in Brosnan, *Jason Lee*, 112.

28. "The Indian's Missionary Address," *Boston Recorder* 24, no. 7 (February 15, 1839): 26; *Christian Advocate and Journal* 14 (October 4, 1839): 25, quoted in Brosnan, *Jason Lee*, 113; *Oregonian and Indian's Advocate* 1 (January 1839): 125–27, quoted in Brosnan, *Jason Lee*, 121; Hines, *Missionary history*, 197; the anecdote also appears in Rev. A. Atwood, *The Conquerors: Historical Sketches of the American Settlement of the Oregon Country* (Cincinnati: Jennings and Graham, 1907), 73n. On staring, see Timothy Mitchell, *Colonising Egypt* (Berkeley: University of California Press, 1991 [1988]), 4–5; and Rosemarie Garland-Thomson, *Staring: How We Look* (New York: Oxford University Press, 2009).

29. *"Ten Thousand Chinese Things": A Descriptive Catalogue of the Chinese Collection in Philadelphia with Miscellaneous Remarks upon the Manners and Customs, Trade and Government of the Celestial Empire* (Philadelphia: Printed for the Proprietor, 1839); Combe, *Notes*, 1:189–90; James Silk Buckingham, *Eastern and Western States*, 3 vols. (London: Fisher, Son & Co., [1842]), 2:55, 64; "Chinese Museum," *Niles National Register*, February 16, 1839, 5, abridges a report on the collection that appeared in Benjamin Silliman's *American Journal of Science and the Arts* in January 1839.

30. Jonathan Goldstein, *Philadelphia and the China Trade, 1682–1846: Commercial, Cultural, and Attitudinal Effects* (University Park: Pennsylvania State University Press, 1978), 50–78.

31. *"Ten Thousand Chinese Things,"* 15–16.

32. Morton, *Crania Americana*, 206.

33. Daniel Wilson, *On the Supposed Prevalence of One Cranial Type throughout the American Aborigines* (Edinburgh: Neill and Company, 1858), 19.

34. Morton, *Crania Americana*, 206.

35. G. Brown Goode, "The Genesis of the National Museum," *Annual Report of the Board of Regents of the Smithsonian Institution* (Washington, DC: Government

Printing Office, 1891), 304, 349; William Baird is quoted in Douglas E. Evelyn, "The National Gallery at the Patent Office," in *Magnificent Voyagers: The U.S. Exploring Expedition, 1838–1842*, ed. Herman J. Viola and Carolyn Margolis (Washington, DC: Smithsonian Institution Press, 1985), 234.

36. Townsend, *Narrative*, 256.

37. Morton, *Crania Americana*, 206, 213; mention of Townsend's reeking pack appears in Patterson, "Memoir," xxix; John Kirk Townsend to Morton, September 20, 1835, Morton Papers, APS. Edward Belcher, *Narrative of a Voyage Round the World, Performed in Her Majesty's Ship Sulphur during the Years 1836–1842*, 2 vols. (London: Henry Colburn, Publisher, 1843), 1:292; Charles Wilkes, *Narrative of the United States Exploring Expedition*, 5 vols. (Upper Saddle River, NJ: Gregg Press, 1970 [1845]), 4:368, 389.

38. Townsend, *Narrative*, 176. Matthias Weaver, an Ohio-born artist who sketched skulls for Morton, had a low opinion of Townsend's book. He described him as a "bold sort of a chap but not much of a writer." Diary of Matthias Weaver, May 21, 1840, microfilm, Ohio Historical Society.

39. Wilson, "Illustrations," 25.

40. Townsend, *Narrative*, 176–77. The editors of the *Family Magazine* reprinted this account as "Flathead Indians," 3 (1835–36): 451.

41. Townsend to Morton, September 20, 1835, Morton Papers, APS.

42. Wilkes, *Narrative*, 4:368.

43. Townsend, *Narrative*, 232.

44. Ibid., 181.

45. Ibid., 236–37; see also 180–81.

46. Morton, *Crania Americana*, 207–15; J. A. Meigs, *Catalogue*, 54–55, 57, 63–64; "Receipt Book for Crania Americana," Manuscripts Division, Department of Rare Books and Special Collections, Princeton University Library.

47. McKenney and Hall, *The Indian Tribes of North America*, 2:158; *New York Herald*, May 30, 1839.

48. Hines, *Missionary history*, 196–97.

49. Franz Boas, "The Doctrine of Souls and of Disease among the Chinook Indians," *Journal of American Folklore* 6, no 20 (Jan.–Mar. 1893): 39–43.

50. Lee, "Some Farther Account of Wm. Brooks"; Whaley, "'Trophies' for God," paras. 17–23.

51. *North American*, quoted in Brosnan, *Jason Lee*, 140–41n.

52. Ibid.

53. Combe, *Notes*, 2:48–49.

54. "Oregon Indians," *New-York Spectator*, November 15, 1838.

55. Combe, *Notes*, 2:49–50.

56. David Meredith Reese, *Humbugs of New York: Being a Remonstrance against Popular Delusions* (New York: J.S. Taylor, 1838), 64, 69; "Dr. Buchanan in Florida," *American Phrenological Journal*, December 1, 1839, 2, 3.

57. Combe, *Notes*, 2:50n.

58. Bruno Latour puts it this way: "In the eyes of our critics the ozone hole above our heads, the moral law in our hearts, the autonomous text, may each

be of interest, but only separately. That a delicate shuttle should have woven together the heavens, industry, texts, souls and moral law—this remains uncanny, unthinkable, unseemly." *We Have Never Been Modern*, trans. Catherine Porter (Cambridge, MA: Harvard University Press, 1993 [1991]), 5.

59. Mark S. Schantz, *Awaiting the Heavenly Country: The Civil War and America's Culture of Death* (Ithaca, NY: Cornell University Press, 2008); Lewis O. Saum, "Death in the Popular Mind of Pre–Civil War America," in *Death in America*, ed. David E. Stannard (Philadelphia: University of Pennsylvania Press, 1975), 30–48.

60. Whaley, "'Trophies' for God," para. 32.

CHAPTER 3

1. Stephens, *Science, Race, and Religion*, 1–10, 19–20; John W. Freeman, *Stories of the Great Operas* (New York: Norton, 1984), 158.

2. *Southern Patriot*, April 16, 1839, 3.

3. Morton, *Crania Americana*, 292.

4. Amos Dean to Samuel George Morton, November 27, 1844, Morton Papers, LCP; "Crania Americana," *Journal of Belles Lettres*, 9; "Letters of Samuel Forry, Surgeon U.S. Army, 1837–1838. Part III," *Florida Historical Quarterly* 7, no. 1 (July 1928): 99.

5. Morton, *Crania Americana*, 5, 6.

6. Ibid., 3, 229; Samuel George Morton to E. G. Squier, April 10, 1847, in Stanton, *The Leopard's Spots*, 84; Morton, *Inquiry into the Distinctive Characteristics*; E. G. Squier and E. H. Davis, *Ancient Monuments of the Mississippi Valley; comprising the results of Extensive Original Surveys and Explorations* (New York: Bartlett & Welford, 1848). See also Robert Silverburg, *The Mound Builders* (Athens: Ohio University Press, 1968); and Gordon M. Sayre, "The Mound Builders and the Imagination of American Antiquity in Jefferson, Bartram, and Chateaubriand," *Early American Literature* 33 (1998): 225–49.

7. Morton, "Observations on the size of the Brain," 223n.

8. Samuel George Morton, "Hybridity in Animals, considered in reference to the question of the Unity of the Human Species," *American Journal of Science*, 2nd ser., 3 (1847): 39, 50, 203–12; Stephens, *Science, Race, and Religion*, 165–94; John S. Haller Jr., "The Species Problem: Nineteenth-Century Concepts of Racial Inferiority in the Origin of Man Controversy," *American Anthropologist* 72 (1970): 1319–29.

9. Samuel George Morton, *Crania Ægyptiaca; or, Observations on Egyptian Ethnography, derived from anatomy, history, and the monuments* (Philadelphia: J. Pennington, 1844). The copy with Morton's annotations is in the Department of Rare Books and Special Collections, Princeton University Library. J. W. Bailey to Morton, April 1, 1847, Morton Papers, courtesy of the Library Company of Philadelphia, letter housed at the Historical Society of Pennsylvania.

10. Samuel George Morton to J. E. Doornik, July 10, 1835, Morton Letter Book; J. E. Doornik to Morton, June 23, 1835, Morton Papers, APS. Phrenologist William Byrd Powell thought of touring Europe with four or five hundred Indian crania. He told Morton that *Crania Americana* anticipated his "purpose as soon

as I could complete my Cabinet to publish a Phrenological view of Savage Crania with plates." Powell to Morton, August 6, 1838, and August 12, 1839, Morton Papers, APS.

11. Matthias Shirk Weaver Diary, July 8, 10, 20, 1841; February 20, 1841; January 18, 1842; February 15, 1843; and June 5, 1843, Ohio Historical Society (microfilm edition). Morton boasted that his fussiness made for better illustrations. "Many of the plates have been drawn the second and third time; and in several instances the entire edition has been cancelled, in order to correct inaccuracies which had previously escaped observation," he wrote. Jeffers Wyman, "Morton's *Crania Americana*," *North American Review* 51, no. 1 (1840): 173. See also Nicholas B. Wainwright, *Philadelphia in the Romantic Age of Lithography* (Philadelphia: Historical Society of Pennsylvania, 1958), 49, 81–82.

12. See Peter Galison, "Judgment against Objectivity," in *Picturing Science, Producing Art*, ed. Caroline A. Jones and Peter Galison (New York: Routledge, 1998), 327–59; Lorraine Daston and Peter Galison, "The Image of Objectivity," *Representations* 40 (Autumn 1992): 81–128; and Samuel Otter, *Melville's Anatomies* (Berkeley: University of California Press, 1999), 110–14.

13. Patterson, "Memoir," xxxiv. On subscription publishing, see Scott E. Casper, "Other Variations on the Trade," in *A History of the Book in America*, vol. 3, *The Industrial Book, 1840–1880*, ed. Scott E. Casper, Jeffrey D. Groves, Stephen W. Nissenbaum, and Michael Winship (Chapel Hill: University of North Carolina, Press, 2007), 219–23. "Receipt Book for Crania Americana," "Crania Americana," *Journal of Belles Lettres*, 9. A notice in the *American Phrenological Journal* reminded readers that the book was available for "subscribers only." "Crania Americana," *American Phrenological Journal* 2, no. 3 (December 1, 1839): 143.

14. Combe to Morton, January 13, 1840, Morton Papers, APS. George Engleman wrote from Germany to ask about a rumor that Morton planned to sell the skulls. Engleman to Morton, February 14, 1841, Morton Papers, APS.

15. Books designed for the British market carried a second dedication to James Cowles Prichard (1786–1848), the naturalist who pioneered the study of man that came to be called "ethnology." Morton, *Crania Americana* (London: Simpkin and Marshall, 1839). On Prichard, see George W. Stocking Jr., "From Chronology to Ethnology: James Cowles Prichard and British Anthropology 1800–1850," in James Cowles Prichard, *Researches into the Physical History of Man* (Chicago: University of Chicago Press, 1973 [1813]), xxiii–cx. Combe objected to the dedication, pointing out that Prichard was no friend to phrenology. Combe to Morton, March 13 and May 13, 1840, Morton Papers, APS. Morton dedicated the American edition to his collaborator John S. Phillips and to his friend the naturalist W. S. W. Ruschenberger. The dedication pleased Ruschenberger: "For the first time in my life I find myself presented with a boon so great that I can neither speak nor act my thanks." Ruschenberger to Morton December 10, 1839, Morton Papers, APS.

16. Combe to Morton, August 20, 1840, Morton Papers, APS.

17. Patterson included Humboldt's compliment in his "Memoir," xxxv; Joel Poinsett to Morton, April 1, 1840, Morton Papers, APS; Benjamin Silliman to Morton, May 6, 1840, Morton Papers, APS; John Collins Warren to Morton,

April 24, 1840, Morton Papers, APS; "Certificate of Membership, Société Royale des Antiquerien du Nord," April 16, 1843, Morton Papers, Microfilm, APS; R. R. Haight to Morton, May 11, 1844, Morton Papers, LCP; C. Meigs, *A Memoir*, 24, 45, 48; Diary of Matthias Weaver, March 31, 1840; Aaron Sachs, *The Humboldt Current: Nineteenth-Century Exploration and the Roots of American Environmentalism* (New York: Viking, 2006), 241–42.

18. C. Meigs, "Extracts," 164.

19. Patterson, "Memoir," xviii, xxi. See also Nelson, *National Manhood*, 102–34.

20. George Combe, "Phrenological Remarks on the relation between the natural Talents and Dispositions of Nations, and the Developments of their Brains," in Morton, *Crania Americana*, 269–91; Aleš Hrdlička to Edward J. Nolan, May 2, 1911, Morton Papers, APS.

21. See Davis, *Phrenology, Fad and Science*; Stern, *Heads and Headlines*; Colbert, *A Measure of Perfection*; and Cooter, *The Cultural Meaning of Popular Science*.

22. A. Cameron Grant, "George Combe and American Slavery," *Journal of Negro History* 45, no. 4 (October 1960): 259–69; Combe, *Notes*, 1:158.

23. Combe, *Notes*, 1:175–77, 307–8.

24. Ibid., 1:195, 243, 359, 361. In *The Philadelphia Negro* (Philadelphia: Published for the University, 1899), W. E. B. DuBois mentions Johnson as a popular bandleader (36). Frederick Douglass described for a British audience his efforts to gain "admission into any place of worship, instruction, or amusement . . ." "*We don't allow niggers in here*." *My Bondage and My Freedom* (Urbana: University of Illinois Press, 1987 [1855]), 226–27.

25. On popular entertainments, see Neil Harris, *Humbug: The Art of P.T. Barnum* (Chicago: University of Chicago Press, 1973); Lawrence Levine, *Highbrow/Lowbrow: The Emergence of Cultural Hierarchy in America* (Cambridge, MA: Harvard University Press, 1990); James W. Cook Jr. *The Arts of Deception: Playing with Fraud in the Age of Barnum* (Cambridge, MA: Harvard University Press, 2001); and Eric Lott, *Love and Theft: Blackface Minstrelsy and the American Working Class* (New York: Oxford University Press, 1995).

26. Combe, *Notes*, 1:91–92; "On the Whitening of Bones," *Philadelphia Medical Museum* 4 (1808): xxxvi–xxxviii.

27. Combe, in Morton, *Crania Americana*, 274–75.

28. Combe, *Notes*, 1:307, 2:186–87; Combe to Morton, March 19, 1839, Morton Papers, APS. Morton appreciated Combe's help, he said, because the man he had asked to write an essay on phrenology had gotten sick and left Philadelphia to "seek a distant and more genial climate." Morton, *Crania Americana*, iii; Harriett Martineau, *Biographical Sketches* (New York: John B. Alden Publishers, 1868), 140; Horace Mann, *A Few Thoughts for a Young Man: A Lecture Delivered before the Boston Mercantile Library Association on Its Twenty-Ninth Anniversary* (Boston: Horace B. Fuller 1871 [1850]), 83. On Combe's readers, see Cooter, *The Cultural Meaning of Popular Science*, 120; James A. Secord, *Victorian Sensation: The Extraordinary Publication, Reception, and Secret Authorship of Vestiges of the Natural History of Creation* (Chicago: University of Chicago Press, 2000), 41–76; and Stack, *Queen Victoria's Skull*, 79–93.

29. Combe, *Notes*, 1:212; Combe to Morton, January 30, March 13, May 31, June 24, and July 18, 1840, Morton Papers, APS; Morton, *Crania Americana*, 262–68; C. Meigs, *A Memoir*, 20, 24.

30. "Review of Morton's *Crania Americana*," *American Journal of Science and Arts* 38 (1840): 341–75; "Crania Americana," *Phrenological Journal*, 143–44; Silliman to Morton, March 27, 1840, Morton Papers, APS.

31. John C. Greene, "Protestantism, Science and American Enterprise: Benjamin Silliman's Moral Universe," in *Benjamin Silliman and His Circle: Studies on the Influence of Benjamin Silliman on Science in America*, ed. Leonard G. Wilson (New York: Science History Publications, 1979), 11–28. A few historians assumed Silliman fell for phrenology. They take Combe's review for an editorial endorsement, just as Combe hoped they would. Davis, *Phrenology, Fad and Science*, cites Silliman's endorsement as evidence that even eminent men accepted phrenology (147). See also Horsman, *Race and Manifest Destiny*, 127. Silliman was busy that spring preparing his Lowell Lectures. See Bruce, *The Launching of Modern American Science*, 12–13, 44–45; and Daniels, *American Science*, 223.

32. Combe to Morton, January 13, March 6, June 24, 1840, and December 23, 1839, Morton Papers, APS.

33. Charles Hodge, "Diversity of Species in the Human Race," *Princeton Review* 34 (July 1862): 444.

34. America's taste for Sir Walter Scott crossed the color line. In 1838 Frederick Douglass took a new last name from the heroic knight in Scott's "Lady in the Lake." Frederick Douglass, *Narrative of the Life of Frederick Douglass, an American Slave*, ed. Robert G. O'Meally (New York: Spark Educational Publishing, 2005 [1845]), ix.

35. Combe, *Notes*, 1:243.

36. *A Statistical Inquiry into the Condition of the People of Color of the City and Districts of Philadelphia* (Philadelphia: Kite & Walton, 1849), 34; Emma Jones Lapansky, "'Since They Got Those Separate Churches': Afro-Americans and Racism in Jacksonian Philadelphia," *American Quarterly* 32 (Spring 1980): 54–78; Julie Winch, *Philadelphia's Black Elite: Activism, Accommodation, and the Struggle for Autonomy, 1787–1848* (Philadelphia: Temple University Press, 1988); Julie Winch, introduction to Joseph Wilson's *Sketches of the Higher Classes of Colored Society in Philadelphia*, ed. Julie Winch (University Park: Pennsylvania State University Press, 2000 [1841]); Gary Nash, *Forging Freedom: The Formation of Philadelphia's Black Community, 1720–1840* (Cambridge, MA: Harvard University Press, 1988).

37. DuBois, *The Philadelphia Negro*, 28; Lapansky, "'Since They Got Those Separate Churches,'" 75.

38. *A. M'Elroy's Philadelphia Directory* (Philadelphia: A. M'Elroy, 1839), 181; Edward Raymond Turner, *The Negro in Pennsylvania: Slavery, Servitude, Freedom, 1639–1881* (New York: Negro Universities Press, 1911), 169–93; Pennsylvania Hall Association, *A History of Pennsylvania Hall, which was destroyed by a Mob, 17th of May 1838* (Philadelphia: Merrihew & Gunn, 1838), 6, 94; Ira V. Brown, "Racism and Sexism: The Case of Pennsylvania Hall," *Phylon* 37, no. 2 (1976): 126–36.

39. James Silk Buckingham, *America, Historical, Statistic, and Descriptive*, 2 vols. (London: Fisher, Son & Co., 1841), 2:96–97.

40. Combe, *Notes*, 2:23.

41. Ibid., 1:265–66.

42. Patterson, "Memoir," lii. On the fate of Combe's corpse, including the study of his brain and head, see Stack, *Queen Victoria's Skull*, 2–5.

43. Valentine Mott, *Travels in Europe and the East* (London: Longman, Brown, Green, and Longman, 1842), 313; "Death of George Gliddon, Esq., Formerly United States Consul in Egypt," *New York Herald Tribune*, November 30, 1857, 5.

44. His arguments with Mohammed Ali, a "tobacco dealer," he called him, appear in two pamphlets Gliddon published in 1841—*A Memoir on the Cotton of Egypt* (London: James Madden & Co., 1841); and *An Appeal to the Antiquaries of Europe on the Destruction of the Monuments of Egypt* (London: James Madden & Co., 1841). According to Gliddon, Egypt's fellahin bore the costs of Ali's agricultural projects. Gliddon also said that Ali had destroyed ancient temples, stripping sandstone blocks for his saltpeter factories. Ali's modernizations damaged Egypt's past and its present.

45. Stanton, *The Leopard's Spots*, 45–46; Patterson, "Memoir," xxxv–xxxvi; "Mummy Cloth," *American Mail*, July 3, 1847; clipping in George R. Gliddon Papers, New-York Historical Society.

46. Gliddon to Morton, March 21, 1839, Morton Papers, APS.

47. Gliddon to Morton, March 21, 1839, May 21–24, 1840, and July 5, 1841, Morton Papers, APS.

48. Morton, *Crania Ægyptiaca*, 1, 66; George R. Gliddon, *Hand-book to the American Panorama of the Nile: Being the original transparent picture exhibited in London, at Egyptian Hall, Piccadilly, purchased from its painters and proprietors, Messrs. H. Warren, J, Bonomi, and J. Fahey* (London: James Madden, 1849); George R. Gliddon, *Ancient Egypt: Her monuments, hieroglyphics, history and archaeology, and other subjects connected with hieroglyphical literature* (Philadelphia: T.B. Peterson, 1850); J. A. Meigs, *Catalogue*, 35–44.

49. Richard Kern headed west in the 1850s, working as an artist and topographer on expeditions led by John C. Frémont and Lt. John Gunnison. In July 1850 he wrote to Morton from Santa Fe, apologizing that he had not been able to send him the unusually large head of Narbona, the elderly Navajo leader who had died. Kern sketched the man. A few years later, Philadelphia naturalists worried about Kern's body, when aggrieved "Parvain Indians" in southern Utah killed Kern and six other men in the Pacific Railroad survey party. California-bound immigrants had attacked native peoples in early summer 1853, disturbing the fragile truce that held between Mormon settlers and native neighbors. Kern's death inspired one of Morton's eulogists to pause to give "his memory a sigh" and to complain that the "bones of Dick Kern bleach unavenged upon the arid plains of Deseret." Patterson, "Memoir," xxxviii–xxxix. It was a sad story all around. Kern's Philadelphia friends would have learned from a Mormon burial party that hungry wolves had cleaned the flesh from the human bones. David J. Weber, *Richard H. Kern: Expeditionary Artist in the Far Southwest, 1848–1853* (Fort Worth: Amon Carter Museum, 1985), 96, 237–43. See also Ned Blackhawk, *Violence over the*

Land: Indians and Empires in the Early American West (Cambridge, MA: Harvard University Press, 2006), 232–36.

50. *Proceedings of the American Philosophical Society, Held at Philadelphia, for the Promoting Useful Knowledge*, 3 (May 25–30, 1843): 115–18; Weaver Diary, September 18, 1843; Fredrickson, *The Black Image in the White Mind*, 77; Lydia Maria Child, *An Appeal in Favor of that Class of Americans Called Africans* (Amherst: University of Massachusetts Press, 1996 [1834]).

51. John A. Wilson, *Signs and Wonders upon Pharaoh* (Chicago: University of Chicago Press, 1964), 41–43; Nelson, "'No Cold or Empty Heart,'" 29–56; Scott Trafton, *Egypt Land: Race and Nineteenth-Century American Egyptomania* (Durham, NC: Duke University Press, 2004), 41–84.

52. George R. Gliddon, *Otia Ægyptiaca: Discourses on Egyptian Archaeology and Hieroglyphic Description* (London: James Madden, 1849), 4–5; Patterson, "Memoir," xxxv–xlii; "Death of George Gliddon," 5; "Circular: Series of Lectures on Early Egyptian History, Archaeology . . . ," Samuel George Morton Papers, Microfilm, APS; Stanton, *The Leopard's Spots*, 47–49.

53. Bruce, *The Launching of Modern American Science*, 41; Stanton, *The Leopard's Spots*, 50.

54. James Ewing Cooley, *The American in Egypt with Rambles through Arabia, Petraea, and the Holy Land during the Years 1839–1840* (New York: D. Appleton & Company, 1842), 364–67; "Mr. Cooley's Remarkable and Amusing Work on Egypt and Egyptians," *New York Herald*, August 6, 1842; "Egyptian Antiquities," *New York Herald*, June 26, 1844; "General Sessions," *New York Herald*, October 10, 1844.

55. Pringle, *The Mummy Congress*, 183–87; "Opening of the Mummy," *Boston Daily Affair*, June 6, 1850; *(Bangor, ME) Daily Whig and Courier*, June 7, 1850; *Cleveland Herald*, June 10, 1850; "Mr. Gliddon's Mummies," *North American and United States Gazette*, June 16, 1851.

56. Edgar Allan Poe, "Some Words with a Mummy" *American Review: A Whig Journal of Politics, Literature, Art, and Science*, April 1845, 363–70. See also Nelson, *National Manhood*, 206–7.

57. Morton, *Crania Ægyptiaca*, 66.

58. "Death of George Gliddon," 5.

59. Edward Lurie, *Louis Agassiz: A Life in Science* (Chicago: University of Chicago Press, 1960), 122; Bruce, *The Launching of Modern American Science*, 30; Louis Menand, *The Metaphysical Club* (New York: Farrar, Straus and Giroux, 2001), 102–12; Elizabeth Cary Agassiz, *Louis Agassiz: His Life and Correspondence* (Boston: Houghton Mifflin, 1886), 419–20, 411; Walls, *Emerson's Life in Science*, 3.

60. Walls, "Textbooks and Texts from the Brooks," 16; Stanton, *The Leopard's Spots*, 102–3; James Wynne, "Louis Agassiz," *Harper's Magazine*, July 1862, 194–201.

61. Jules Marcou, *Life, Letters, and Works of Louis Agassiz*, 2 vols. (New York: Macmillan, 1896), 1:292.

62. Ibid., 1:285–86, 293; E. C. Agassiz, *Louis Agassiz*, 417, 437; Lurie, *Louis Agassiz*, 125; Stanton, *The Leopard's Spots*, 102–3. This letter figures in Menand's illuminating discussion of Agassiz in *The Metaphysical Club*, 105.

63. Patterson, "Memoir," xxxviii, xl. On naturalists' boasts, see Burnett, *Trying Leviathan*, 2.

64. Joseph Willson, *Sketches of the Higher Classes of Colored Society in Philadelphia by a Southerner* (Philadelphia: Merrihew and Thompson, Printers, 1841), 13, 29.

65. Ibid., 64, 28.

66. Morton, *Brief remarks on the diversities of the human species.*

67. Frederick Douglass, *The Claims of the Negro, ethnologically considered: An address before the literary societies of Western Reserve College, at commencement, July 12, 1854* (Rochester, NY: Lee, Mann & Co., Daily American Office, 1854), 5–6; Bay, *The White Image*, 67–71; Audrey McCluskey and John McCluskey, "Frederick Douglass on Ethnology: A Commencement Address at Western Reserve College, 1854," *Negro History Bulletin* 40 (July–August 1977): 747–49.

68. Douglass, *The Claims of the Negro*, 20, 21.

69. Ibid., 14.

CHAPTER 4

1. Charles Pickering to Morton, June 10, 1842, Morton Papers, APS. On May 5, 1836, the *Lynn (MA) Record* reported that Birilip, a "young native of Fegee Islands, brought to this country by Capt. Eagleston, in the ship Emerald, of Salem, is being exhibited at the Baltimore Museum." In R. Gerard Ward, ed., *American Activities in the Central Pacific, 1790–1870*, 8 vols. (Ridgewood, NJ: Gregg Press, 1967), 2:392. On August 4, 1840, Pickering mentions meeting a "Feejee dwarf" at Rewa who had been brought to Salem by Eagleston. Pickering had once seen a native of the "Green Islands" exhibited by a Captain Morrell. "These we believe are the only individuals of this race, that have ever reached America; and we are not aware that hitherto any have been carried to Europe. We once asked a Feejee man why he was unwilling to go to America; and his reply was 'that he was fearful of getting into difficulty about the Women'!" Charles Pickering, Journal, 1838–1841, August 4, 1840, 49, Microfilm, Massachusetts Historical Society.

2. Anthropologist William Henry Flower believed that the skull of a whaler's crewman who died in a hospital at Hobart Town was the only cranium from Fiji in early nineteenth-century European collections. "On the Cranial Characteristics of the Natives of the Fiji Islands," *Journal of the Anthropological Institute of Great Britain and Ireland* 10 (1881): 154; John Collins Warren, *A Comparative View of the Sensorial and Nervous System in Men and Animals* (Boston: J.W. Ingraham, 1822), 93–94. The "races" of the Pacific confused Pickering. See Charles Pickering, *The Races of Man and Their Geographical Distribution* (London: H.G. Bohn, 1850), 147–49. See also Horatio Hale, *United States Exploring Expedition: Ethnography and Philology* (Ridgewood, NJ: Gregg Press, 1968 [1846]), 48–49. Canadian ethnologist Sir Daniel Wilson suggested that Fijian and Egyptian skulls came closest to retaining a "natural shape" because Fijian and Egyptian children learned to rest with heads supported by neck pillows. "Illustrations of the Significance," 27–30. Wilson also argued that culture (simple things like how and how long a woman nursed a baby), not biology, set the shape of the skull. Daniel Wilson, *Ethnical*

Forms and Undesigned Artificial Distortions of the Human Cranium (Toronto: Printed by Lovell and Gibson, 1862). See also Bruce Trigger, "*Prehistoric Man*, and Daniel Wilson's Later Canadian Ethnology," in *Thinking with Both Hands: Sir Daniel Wilson in the Old World and the New*, ed. Marinell Ash (Toronto: University of Toronto Press, 1999), 81–100.

3. Wilkes, *Narrative*, 1:xxv, 5:486–87. Nathaniel Philbrick, *Sea of Glory: America's Voyage of Discovery, the U.S. Exploring Expedition, 1838–1842* (New York: Viking, 2003), gives a wonderful account of the U.S. Ex. Ex. See also William Stanton, *The Great United States Exploring Expedition of 1838–1842* (Berkeley: University of California Press, 1975); Viola and Margolis, *Magnificent Voyagers*; and the materials at the Smithsonian website "The United States Exploring Expedition, 1838–1842," http://www.sil.si.edu/digitalcollections/usexex/usexex_resources.cfm. For a general picture of Pacific exchange, see David Igler, "Diseased Goods: Global Exchanges in the Eastern Pacific Basin, 1770–1850," *American Historical Review* 109 (June 2004): 693–719.

4. J. K. Paulding, "Instructions," in Wilkes, *Narrative*, 1:xxix.

5. Wilkes, *Narrative*, 3:73; William Reynolds, *The Private Journal of William Reynolds: United States Exploring Expedition, 1838–1842*, ed. Nathaniel Philbrick and Thomas Philbrick (New York: Penguin Books, 2004), 144; Morton, *Crania Americana*, 61–62; Pickering, *Races*, 169. On the lingering power of these racial insults, see Epeli Hau'ofa, "Our Sea of Islands," in *Inside Out: Literature, Cultural Politics, and Identity in the New Pacific*, ed. Vilsoni Hereniko and Rob Wilson (Lanham: Rowman & Littlefield, 1999), 28.

6. Wilkes, *Narrative*, 3:221, 316. The quotation from James Dwight Dana is in Barry Alan Joyce, *The Shaping of American Ethnography: The Wilkes Exploring Expedition, 1838–1842* (Lincoln: University of Nebraska Press, 2001), 102. In *Typee: A Peep at Polynesian Life* (1846), Melville toys with the Pacific's promise of romance and its threat of cannibalism.

7. On Mott, see Samuel D. Gross, *Memoir of Valentine Mott, M.D, LL.D., professor of surgery in the University of the city of New York* (New York: D. Appleton and Co., 1868). Diana diZerega Wall and Anne-Marie Cantwell discuss Veidovi's grave in *Touring Gotham's Archaeological Past* (New Haven, CT: Yale University Press, 2004), 166–67.

8. "Important News from the Exploring Expedition," *Weekly Herald*, June 18, 1842; Jesse Poesch, *Titian Ramsay Peale, 1799–1885, and His Journals of the Wilkes Expedition* (Philadelphia: American Philosophical Society, 1961), 84.

9. "LAST OF VENDOVI!," *New York Herald*, June 12, 1842, 1. See also P. Turnbull, "'Rare Work,'" 4.

10. William H. Goetzmann, *New Lands, New Men: America and the Second Great Age of Discovery* (New York: Viking, 1986), 289. On U.S. Ex. Ex. collections, see Adrienne L. Kaeppler, "Anthropology and the U.S. Exploring Expedition," in *Magnificent Voyagers*, ed. Viola and Margolis, 119–48. Peale complained of one lieutenant's "contempt for the operations of the Scientific Corps." Poesch, *Titian Ramsay Peale*, 176.

11. Hau'ofa, "Our Sea of Islands," 33.

12. Joyce, *Shaping of American Ethnography*, 155.

13. George Hamilton, "Biographical Sketch of James Aitken Meigs, M.D.," *Transactions of the Medical Society of the State of Pennsylvania for 1880* (Philadelphia: Collins, Printer, 1880), 6–17; Laurence Turnbull, "Memoir of James Aitken Meigs, M.D.," *Medical Bulletin* 3, no. 2 (February 1881): 33–37; James Aitken Meigs, "The Cranial Characteristics of the Races of Men," in *Indigenous Races of the Earth*, ed. Josiah C. Nott and George R. Gliddon (Philadelphia: J.B. Lippincott & Co., 1857), 213.

14. J. A. Meigs, *Hints to craniographers*, 7–8.

15. Ibid., 7; Alfred Hunter, *A Popular Catalogue of the Extraordinary Curiosities in the National Institute arranged in the Building belonging to the Patent Office* (Washington, DC: Published by Alfred Hunter, 1855). See Barbara Kirshenblatt-Gimblett, "Objects of Ethnography," in *Exhibiting Cultures: The Poetics and Politics of Museum Display*, ed. Ivan Karp and Steven D. Lavine (Washington, DC: Smithsonian Institution Press, 1991), 386–443; and Tony Bennett, *The Birth of the Museum: History, Theory, Politics* (London: Routledge, 1995).

16. Goode, "The Genesis of the National Museum," 285, 318, 321; Evelyn, "The National Gallery," 236–38; Charles Wilkes, *Autobiography of Rear Admiral Charles Wilkes, U.S. Navy, 1798–1877*, ed. William James Morgan, David B. Tyler, Joye L. Leonhart, and Mary F. Loughlin (Washington, DC: Naval History Division, 1978), 529. See also *Objects in the National Gallery, Patent Office Building; as Originally Arranged by Dr. Chas. Pickering, T. R. Peale, James D. Dana, and Others of the United States Exploring Expedition* (Washington, DC: Henry Polkinhorn, Printer, 1856). The Fijian war club inspired one visitor to meditate on the century's escalating violence. "In the ratio of our civilization, with a religion of peace, and exalted worship—with moral as well as intellectual elevation, behold the land strewed with weapons of death! Observe our fierce spirit of contest—our arsenals stored with destruction, as yet silent and unscattered; but ready for their grim interpreters and disseminators: the war ship, the mine and red artillery. We invoke battle and invite famine—its gaunt handmaiden—to our realms. May not the Being whose intelligence levels all earthly superiority, consider our wrath and acts of bloodshed to be as inexcusable and preposterous, as we estimate them to be in the savage of those lovely islands." Hunter, *A Popular Catalogue*, 19.

17. Kaeppler, "Anthropology," 121; J. A. Meigs, *Catalogue*, 55. The reference to the trade in heads is in Hunter, *A Popular Catalogue*, 49–50. Captain David Porter describes islanders "with skulls to traffic for harpoons" in his *Journal of a Cruise Made to the Pacific Ocean*, 2 vols. (New York: Wiley & Halsted, 1822), 2:115. Harvard doctor John Collins Warren also describes purchase of a Pacific Island skull in *Comparative View*, 142.

18. Hunter, *A Popular Catalogue*, 49–50. Accounts of the eye-eating Fijian appear in Simeon A. Stearns, "Journal on Board U.S.S. Vincennes of the so-called Wilkes Exploring Expedition to the South Pacific: 13 August 1839–5 September 1840," 75, Simeon A. Stearns Papers, Manuscripts and Archives Division, the New York Public Library, Astor, Lenox and Tilden Foundations. Wilkes describes purser William Spieden buying the head in Wilkes, *Narrative*, 3:234.

19. R. A. Derrick, *A History of Fiji* (Suva, Fiji: Printing and Stationery Department, 1946), 38–39.

20. George Foster Emmons, Journal, June 7, 1840, 3 vols., Yale Collection of Western Americana, Beinecke Rare Book and Manuscript Library, Yale University.

21. H. E. Maude, "Beachcombers and Castaways," *Journal of the Polynesian Society* 73 (1964): 254–93; Marshall Sahlins, "The Discovery of the True Savage," in *Culture in Practice: Selected Essays* (New York: Zone Books, 2000), 375–80; Marshall Sahlins, *Apologies to Thucydides: Understanding History as Culture* (Chicago: University of Chicago Press, 2004), 30–38; G. C. Henderson, *Fiji and the Fijians, 1835–1856* (Sydney: Angus & Robertson, 1931), 31; Wilkes, *Narrative*, 3:323; Derrick, *History*, 37.

22. Wilkes, *Narrative*, 3:323; John S. Jenkins, *Voyage of the U.S. Exploring Squadron, commanded by Captain Charles Wilkes of the United States Navy* (Auburn, NY: James Alden, 1850), 332; James Oliver, *The Wreck of the Glide with Recollections of the Fijis and of Wallis Island* (New York: Wiley and Putnam, 1848), 47–48.

23. "Journal of William L. Hudson, comdg. U.S. Ship *Peacock*, one of the vessels attached to the South Sea Surveying and Exploring Expedition under the command of Charles Wilkes Esq.," 1:498, American Museum of Natural History Archives; Wilkes, *Narrative*, 2:73; Reynolds, *Private Journal*, 144–45, 150; William Reynolds, *Voyage to the Southern Ocean: The Letters of Lieutenant William Reynolds from the U.S. Exploring Expedition, 1838–1842*, ed. Anne Hoffman Cleaver and E. Jeffrey Stann (Annapolis: Naval Institute Press, 1988), 169; Belcher, *Narrative*, 2:50–51; Pickering, Journal, 154. "Dissolute and besotted" Phillips died at age forty, a victim of dysentery brought on by poisoning from his home-distilled grog of bananas, sugarcane, and "the root of the wild dracaena." See Derrick, *History*, 105; John Jackson, "Jackson's Narrative," in John Elephinstone Erskine, *Journal of a Cruise among the Islands of the Western Pacific including the Feejees* (London: J. Murray, 1853), 461–63.

24. Wilkes, *Narrative*, 3:77–78, 289; Pickering, Journal, 154–55; Marshall Sahlins, *Islands of History* (Chicago: University of Chicago Press, 1985), 75–103; Sahlins, *Apologies*, 221.

25. Wilkes, *Narrative*, 3:73–105; Joseph Waterhouse, *The King and People of Fiji: Containing a Life of Thakombau: with notices of the Fijians, their manners, customs, and superstitions, previous to the great religious reformation in 1854* (London: Wesleyan Conference Office, 1866), 44–46; Johannes Fabian, *Time and the Other: How Anthropology Makes Its Object* (New York: Columbia University Press, 1983).

26. Hale, *United States Exploring Expedition*, 51; Henderson, *Fiji and Fijians*, 244–45; Waterhouse, *The King and People*, 12–13. Marshall Sahlins extends the comparison, describing Bau and Rewa as Athens and Sparta in the Pacific. *Apologies*, 13–124.

27. Derrick, *History*, 39; Everard Im Thurn, ed., *Lockerby's Journal* (London: Hakluyt Society, 1925).

28. R. Gerard Ward, "The Pacific *Bêche-de-Mer* Trade with Special Reference to Fiji," in *Man in the Pacific Islands: Essays on Geographical Change in the Pacific*

Islands, ed. R. Gerard Ward (Oxford: Oxford University Press, 1972), 91–123; Oliver, *Wreck of the Glide*, 35–36; Wilkes, *Narrative*, 3:220–22.

29. Derrick, *History*, 69.

30. William Endicott, *Wrecked among Cannibals in the Fijis: A Narrative of Shipwreck and Adventure in the South Seas* (Salem, MA: Marine Research Society, 1923), 24; Frederick J. Simoons, *Food in China: A Cultural and Historical Inquiry* (Boca Raton, FL: CRC Press, 1991), 435. Though it would have been possible to ship bêche-de-mer to Boston or New York, contemporaries did not imagine that even oyster-loving Americans would go for sea slugs. "There are various modes of preparing them for the table; but as it is not likely they will be ever used in this country, it is needless to give them," wrote one contemporary. Hunter, *A Popular Catalogue*, 12. For the less squeamish, here is a recipe for Braised Bêche-de-Mer with Scallions. 1. Clean soaked sea slug and cook in boiling water for two minutes. Fry scallion pieces in lard until brown. Remove scallions and pour out half the lard. Put in sea slug, chicken soup and simmer until done. 2. Place sea slug on a large tray. Thicken chicken soup with cornstarch. Add two rape plants and pour over sea slug along with scallion oil. Adapted from *Imperial Dishes of China* (Hong Kong; Tai Dao Publishing, 1986), 24.

31. Derrick, *History*, 68–69; Wilkes, *Narrative*, 3:220–22; Oliver, *Wreck of the Glide*, 34–38; J. H. Eagleston, "Journal of the Ship Emerald of Salem, 1833–1836," Essex Institute and Phillips Library, 250–57.

32. Sahlins, "The Discovery of the True Savage," 371–80; Derrick, *History*, 68, 71; Ward, "The Pacific *Bêche-de-Mer* Trade," 109–10; Sahlins, *Apologies*, 34–35; Wilkes, *Narrative*, 3:103; John Coulter, M.D., *Adventures in the Pacific; with observations on the Natural Productions, Manners and Customs of the Natives of the Various Islands; together with remarks on Missionaries, British and Other Residents* (Dublin: William Curry, Jun., and Company, 1845), 177–78.

33. Ward, "The Pacific *Bêche-de-Mer* Trade," 96–99, 102–6.

34. Pickering, Journal, 122; Mary Wallis, *Life in Feejee: Five years among the Cannibals: A Woman's account of Voyaging the Fiji Islands aboard the 'Zotoff' (1844–49)* (Santa Barbara, CA: Narrative Press, 2002 [1851]). Sahlins suggests that Fijian men who saw themselves as warriors were reluctant to turn to fishing. But the remaking of labor and power that grew out of the sea slug trade opened the country to Christian conversion and finally to its incorporation as a protectorate under the Britain's Pacific empire. A second sea slug rush began in the 1840s, and fish supplies held out through the 1850s, attracting new American traders even as Fijians were embroiled in a bloody civil war. Sahlins, *Apologies*, 98–100.

35. Ward, "The Pacific *Bêche-de-Mer* Trade," 91–123; Stearns, Journal, 188; Reynolds, *Private Journal*, 164, 237; Sahlins, "The Discovery of the True Savage," 375–80.

36. Wilkes, *Narrative*, 3:357; "Distressing Outrage," *Lynn (MA) Record*, September 24, 1834, in Ward, *American Activities*, 2:379.

37. Maude, "Beachcombers and Castaways," 261; William S. Cary, *Wrecked on the Feejees* (Fairfield, WA: Ye Galleon Press, 1988); William Diapea, *Cannibal Jack: The True Autobiography of a White Man in the South Seas* (New York: G. P. Putnam's

Sons, 1928); Pickering, *Races*, 146; Joseph G. Clarke, *Lights and Shadows of Sailor Life, as Exemplified in Fifteen Years' Experience* (Boston: Mussey, 1848), 140; William Clark, "Logbook of the ship Vincennes, July 1839–May 1842," used by permission of the Phillips Library at the Peabody Essex Museum, Salem, MA.

38. Wilkes, *Narrative*, 3:361. Monogamous witnesses had a hard time keeping track of the number of Connel's wives. Was there really a difference between five and fifty-five?

39. Stearns, Journal, 188–91; Charles Erskine, *Twenty Years before the Mast* (Boston: By the Author, 1890), 153.

40. This instance of cannibalism moves through the accounts like some version of the children's game of telephone. When Wilkes wrote his autobiography in the 1870s, he reversed the story. "Of a crew of eight only one, a negro, was saved for a time from infernale cannibalism. The rest were all eaten. The reason assigned of their not devouring the Negro cook was that his flesh would taste so much of tobacco." Did taste? Would taste? Wilkes, *Autobiography*, 475. Stearns writes that the "Cook tasted so strong of Tobacco they could not eat him." He adds that the hearts and livers of the others had "been taken and roasted & eaten." Stearns, Journal, 189.

41. Wilkes, *Narrative*, 3:103–5, 361; Reynolds, *Private Journal*, 155–63; Hudson, Journal, 495.

42. Wilkes, *Narrative*, 3:113; Hudson, Journal, 495, 497, 499; Derrick, *History*, 91.

43. Joel Bulu, *The Autobiography of a Native Minister in the South Seas* (London: T. Woolmer, 1884), 35; Belcher, *Narrative*, 2:39, 51; Wilkes, *Narrative*, 3:357–58.

44. Hudson, Journal, 499–500; Reynolds, *Voyage*, 173; Wilkes, *Narrative*, 3:129, 363; Derrick, *History*, 57.

45. Americans had drilled and danced for the Fijians before; "a sight that always pleases a savage," Reynolds wrote. *Voyage*, 167. Reynolds also described an occasion when the marines paraded to the tune of "The King of the Cannibal Islands." *Private Journal*, 148, 156–58. Several witnesses described a performance in which a marching marine fired off his musket mid-drill, barely missing some in the audience. Pickering, Journal, 156; Hudson, Journal, 500; Erskine, *Twenty Years*, 160–61.

46. Wilkes, *Narrative*, 3:118, 120. Talented Agate died of consumption in Washington in 1846, a few weeks shy of his thirty-fifth birthday. Veidovi's arrival fits among the scenes Garland-Thomson describes in *Staring: How We Look*.

47. Hudson, Journal, 500–501; Reynolds, *Private Journal*, 154, 159; Wilkes, *Narrative*, 3:118, 120. Scots-born Cargill, who had devised a Fijian alphabet, translated the trade regulations for Hudson. He came aboard to see "Veindovi . . . in irons," leaving his very pregnant wife home on shore. (She died in childbirth later that month, leaving him with three small daughters. The family buried her body in Fiji.) David, Cargill, *The Diaries and Correspondence of David Cargill, 1832–1843*, ed. Albert J. Schütz (Canberra: Australian National University Press, 1977), 180–81.

48. Reynolds, *Voyage*, 175; Reynolds, *Private Journal*, 158–60; Wilkes, *Narrative*, 3:136–37.

49. Reynolds, *Private Journal*, 161; Wilkes *Narrative*, 3:134; Wallis, *Life in Feejee*, 22; Here is how Missionary Waterhouse enumerated the fates of the sons of the king of Rewa in 1866. "1. Koroitamana, the parricide killed by feudal soldiers. 2. Macanawai, clubbed by his brother Tuisawau. 3. Tuisawau, murdered by gun-shot by his brother Vendovi. 4. Vendovi, given in retaliation by his brothers to Captain Wilkes, and transported to meet an early death. 5. Banuvi (the king) killed by the hands of his Bauan cousins, in war. 6. Vakatawanavatu, killed in war." Waterhouse, *The King and People*, 42.

50. Wilkes, *Narrative*, 3:134–35, 138. George Foster Emmons refers to the barber as "Jack, a Wahoo man," who has accompanied "Bendova" on board. Journal, May 28, 1840.

51. Wilkes, *Narrative*, 3:138; Hale, *United States Exploring Expedition*, 62.

52. Magoun was an Ulster-born blacksmith who had survived the wreck of the *Fawn* off the reefs of Somosomo in 1830. The local people valued his metal working skills, and he lived in Somosomo until Europeans in Levuka arranged a ransom of 12 knives, 6 muskets, 2 rolls of lead, 2 kegs of powder, 1 bundle of hoop iron, 2 bolts of red cloth, 20 pounds of red paint, 2 large iron pots, 20 pounds of beads, and 1 Tongan girl. Donald Akenson, *An Irish History of Civilization* (London: Granta, 2005), 548–49.

53. Wilkes, *Narrative*, 3:103–5, 138, 411–14. On beachcombers' stories, particularly cannibal stories, see Gananath Obeyesekere, "Narratives of the Self: Chevalier Peter Dillon's Fijian Cannibal Adventures," in *Body Trade*, ed. Creed and Hoorn, 69–111. The *New-York Spectator* ran a story on September 25, 1834, reporting that Captain Bachelor had written from Manila on April 7 to report the murder of nine of his men the previous September. The story sets the murders in 1833. Others say 1832. Cargill, *Diaries*, 180. In any case, both Connel and Magoun were mistaken when they described the murders taking place in 1834. In November 1834, the Salem East India Marine Society passed a resolution urging the U.S. government to support a "voyage of discovery and survey to the South seas." Listed among the motives for such a voyage was the loss of the crew of the *Charles Doggett*. News would not have reached the States in two months.

54. Pickering, Journal, 293; George M. Colvocoresses, *Four Years in the Government Exploring Expedition* (New York: R.T. Young, 1853), 150; Wilkes, *Narrative*, 5:539.

55. Wilkes, *Autobiography*, 474–75; Pickering, *Races*, 147.

56. Clarke, *Lights and Shadows*, 154–55; Silas Holmes, Journal, 59, Yale Collection of Western Americana, Beinecke Rare Book and Manuscript Library, Yale University.

57. Wilkes, *Narrative*, 3:267.

58. Ibid., 3:271–73; Reynolds, *Voyage*, 190. According to Silas Holmes, "those incarnate devils have been repeatedly known to dig up and devour the *decomposing bodies of the dead*." Holmes, Journal, 58.

59. Oliver, *Wreck of the Glide*, 178.

60. *Louisville Public Advertiser*, February 17, 1841; Holmes, Journal, 67.

61. Wilkes, *Narrative*, 3:265, 283, 342–43; Reynolds, *Private Journal*, 118; Holmes, Journal, 68–69.

62. Wilkes, *Narrative*, 3:280, 283; Reynolds, *Private Journal*, 237; Karl Jacoby, "'The Broad Platform of Extermination': Nature and Violence in the Nineteenth-Century North American Borderlands," *Journal of Genocide Research* 10, no. 2 (June 2008): 252; Hale, "Vitian Dictionary," in *United States Exploring Expedition*, 392. Pickering on Oahu Sam, Journal, 25. The barber/interpreter did not leave Fiji with the Americans. Wilkes lists a crewman, one of seven men named "John Smith," who "joined at Feejee Islands; discharged at same place." This John Smith is the only man both to join and to leave at Fiji. The mention may mark the barber/interpreter's place on Wilkes's payroll. Wilkes, *Narrative*, 1:lii; Derrick, *History*, 23.

63. Wilkes, *Narrative*, 1:353, 3:265, 286.

64. Ibid., 3:316.

65. Reynolds, *Private Journal*, 164. The exchange is a particularly apt example of the accumulation and transformation of information that Latour describes in *Science in Action*, 215–23.

66. Pickering, *Races*, 15, 40, 167; Wilkes, *Narrative*, 4:297; Erskine, *Twenty Years*, 236.

67. William Heath Davis, *Sixty Years in California: A History of Events and Life in California; Personal, Political and Military, Under the Mexican Regime; During the Quasi-Military Government of the Territory by the United States, and after the Admission of the State into the Union* (San Francisco: A.J. Leary, 1889), 132; Henry M. Lyman, *Hawaiian Yesterdays: Chapters from a Boy's Life in the Islands in the Early Days* (Chicago: A.C. McClurg & Co., 1906), 54–55.

68. Wilkes, *Autobiography*, 475.

69. Clark, "Logbook"; Wilkes, *Narrative*, 5:417–18; Pickering, Journal, 15–16.

70. Clark, "Logbook"; Pickering to Morton, June 10, 1842, Morton Papers, APS.

71. William H. Goetzmann, *Exploration and Empire: The Explorer and the Scientist in the Winning of the American West* (New York: Norton, 1966), 237–38.

72. Wilkes, *Narrative*, 1:367–68; Daniel C. Haskell, *The United States Exploring Expedition, 1838–1842, and Its Publications, 1844–1874* (New York: New York Public Library, 1942); Philbrick, *Sea of Glory*, 344.

73. Stanton, *The Great United States Exploring Expedition*, 257, 266, 275, 278, 280, 281; Evelyn, "The National Gallery," 227–42; Nathan Reingold and Marc Rothenberg, "The Exploring Expedition and the Smithsonian Institution," in *Magnificent Voyagers*, ed. Viola and Margolis, 243–54; Herman J. Viola, "The Story of the U.S. Exploring Expedition," in ibid., 20, 21.

74. "U.S. Exploring Expedition," *Southern Literary Messenger* 2, no. 5 (May 1845): 322; *New York Herald*, June 11, 17, 19, July 26, 28, 1842.

75. "American Museums and Garden," *New York Herald*, June 28, 1842.

76. P. T. Barnum, *Life of P. T. Barnum, Written by Himself* (Urbana: University of Illinois Press, 2000 [1855]), 234–35. On the mermaid's career in England, see Jan Bondeson, *The Fejee Mermaid and Other Essays in Natural and Unnatural History*

(Ithaca, NY: Cornell University Press, 1999), 41–50; and Harriet Ritvo, *The Platypus and the Mermaid and Other Figments of the Classifying Imagination* (Cambridge, MA: Harvard University Press, 1997), 178–82; see also Reiss, *The Showman and the Slave*; Harris, *Humbug*, 62–67; Kenneth Greenberg, *Honor and Slavery* (Princeton, NJ: Princeton University Press, 1996), 3–16; Cook, *The Arts of Deception*, 73–118; and O. S. Fowler, *Self Culture and Perfection of Character, including Management of Youth* (New York: Fowler & Wells, Publishers, 1847), 202–3.

77. Colvocoresses, *Four Years*, 168–69; Jared L. Elliott, "A Sermon Occasioned by the Death of Lieutenant J.A. Underwood and Midshipman Wilkes Henry, of the United States Navy, delivered on board the U.S. Ship Vincennes before the Officers and Men of the U.S. Exploring Expedition, August 10, 1840, by Jared L. Elliott, Chaplain USN" (Honolulu, Oahu: Mission Press, 1840); Wilkes, *Narrative*, 3:311.

78. Historian of Fiji R. A. Derrick assumed Veidovi served "a sentence in America; but nothing more was heard of him, and the lesson was not lost upon other Fijian chiefs." *History*, 56–57, 74–77; Wilkes, *Narrative*, 3:98–99; Wallis, *Life in Feejee*, 75–77; Lorimer Fison, "Notes on Fijian Burial Customs," *Journal of the Anthropological Institute of Great Britain and Ireland* 10 (1881): 137–49.

79. Joyce, *Shaping of American Ethnography*, 155; T. D. Stewart, "The Skull of Vendovi: A Contribution of the Wilkes Expedition to the Physical Anthropology of Fiji," *Archaeology and Physical Anthropology in Oceania* 13, nos. 2 & 3 (July and October 1978), 204–14; Adrienne Kaeppler, "Two Polynesian Repatriation Enigmas at the Smithsonian Institution," *Journal of Museum Ethnography* 17 (2004): 152–62.

80. Rhees, *Account of the Smithsonian Institution*, 69, labels Veidovi a "murderer." Aleš Hrdlička, *Physical Anthropology: Its Scope and Aims, Its History and Present Status in the United States* (Philadelphia: Wistar Institute, 1919), 116.

81. Stewart, "The Skull of Vendovi"; Kaeppler, "Two Polynesian Repatriation Enigmas," 159–60.

82. Brij V. Lal, "Heartbreak Islands: Reflections on Fiji in Transition," in *Law and Empire in the Pacific: Fiji and Hawai'i*, ed. Sally Engle Merry and Donald Brenneis (Santa Fe: School of American Research Press, 2003), 261–80.

CHAPTER 5

1. George McGill to John Shaw Billings, November 14, 1866, John Shaw Billings Papers, New York Public Library, box 1; A. T. McGill to John Shaw Billings, November 1, 1867, box 1, John Shaw Billings Papers, Manuscripts and Archives Division, New York Public Library, Astor, Lenox and Tilden Foundations (hereafter Billings Papers, NYPL).

2. "An Army Medical Museum," *New York Times*, November 15, 1862, 2; Carleton B. Chapman, *Order Out of Chaos: John Shaw Billings and America's Coming of Age* (Boston: Boston Medical Library, 1994), xiii, quotes Billings describing his search for order in a chaotic field hospital at Gettysburg. See also Fielding H. Garrison, M.D., *John Shaw Billings: A Memoir* (New York: G. P. Putnam's Sons, 1915);

and Frank Bradway Rogers, compiler, *Selected Papers of John Shaw Billings* (Baltimore: Medical Library Association, 1965).

3. A. T. McGill to Billings, November 1, 1867, Billings Papers, NYPL; "Died," *New York Times*, July 26, 1867, 5; "The Plains," *New York Times*, August 7, 1867, 8; "Prof. M'Gill Very Ill," *New York Times*, December 5, 1888, 2.

4. Select Committee Relative to the Soldiers' National Cemetery, *Revised Report of the Select Committee Relative to the Soldiers' National Cemetery, together with the accompanying documents, as reported to the House of Representatives of the Commonwealth of Pennsylvania* (Harrisburg, PA: Singerly and Myers, 1865), 7. Several historians have written on civilian struggles to come to terms with the Civil War's death toll. The most insightful is Drew Faust's *The Republic of Suffering*. That study builds on excellent work: Gary Laderman, *The Sacred Remains: American Attitudes toward Death, 1799–1883* (New Haven, CT: Yale University Press, 1996); Garry Wills, *Lincoln at Gettysburg: The Words That Remade America* (New York: Simon and Schuster, 1992), 63–89; Karen Flood, "Contemplating Corpses: The Dead Body in American Culture, 1870–1920" (Ph.D. diss., Harvard University, 2001); and Kirk Savage, *Standing Soldiers/Kneeling Slaves: Race, War, and Monument in Nineteenth-Century America* (Princeton, NJ: Princeton University Press, 1997). On middle-class funerary practices, see Karen Halttunen, *Confidence Men and Painted Women: A Study of Middle-Class Culture in America, 1830–1870* (New Haven, CT: Yale University Press, 1982), 124–52; and Ann Douglas, "Heaven Our Home: Consolation Literature in the Northern United States, 1830–1880," in *Death in America*, ed. Stannard, 49–68.

5. William A. Hammond, "Circular no. 2," reprinted in John H. Brinton, *Personal Memoirs of John H. Brinton* (New York: Neale Publishing, 1914), 180.

6. J. J. Woodward, "The Army Medical Museum," *Lippincott's Magazine of Popular Literature and Science*, March 1871, 233.

7. Brinton, *Personal Memoirs*, 187–88; Army Medical Museum, *Catalogue of the Army Medical Museum* (Washington, DC: Government Printing Office, 1863), 9.

8. Brinton, *Personal Memoirs*, 185–86; *Catalogue of the Army Medical Museum* (1863), 6.

9. *Catalogue of the Army Medical Museum* (1863), 30–31; Army Medical Museum, *Catalogue of the Surgical Section of the United States Army Medical Museum* (Washington, DC: Government Printing Office, 1866), 495.

10. Louis Agassiz to Edwin Stanton, January 20, 1865, in Lurie, *Louis Agassiz*, 338.

11. Spencer Baird to J. K. Barnes, January 24, 1869, National Anthropological Archives, Smithsonian Institution, Washington, DC (hereafter cited as NAA), reel 3, 618.

12. "Army Medical Department," *New York Times*, December 7, 1870, 6; Ian Hacking, *The Taming of Chance* (Cambridge: Cambridge University Press, 1990), 61; Daniel S. Lamb, "A History of the Army Medical Museum, 1862–1917, Compiled from Official Records," ms., National Museum of Health and Medicine (n.d.). See also Theodore M. Porter, *The Rise of Statistical Thinking, 1820–1900*

(Princeton, NJ: Princeton University Press, 1986); and Cohen, *A Calculating People*.

13. Harrison, *The Dominion of the Dead*; Robert Hertz, *Death and the Right Hand*, trans. Rodney Needham and Claudia Needham (Glencoe, IL: Free Press, 1960), 29–86; Thomas Lynch, *Bodies in Motion and at Rest: On Metaphor and Mortality* (New York: Norton, 2000); Peter Metcalf and Richard Huntington, *Celebration of Death: The Anthropology of Mortuary Ritual*, 2nd ed. (New York: Cambridge University Press, 1991). The treatment of Native American dead in the post–Civil War United States is one instance of many in the inequitable treatment of bodies of poor people. See Richardson, *Death, Dissection, and the Destitute*; and Randall H. McGuire, "The Sanctity of the Grave: White Concepts and American Indian Burials," in *Conflict in the Archaeology of Living Traditions*, ed. Robert Layton (London: Routledge, 1989), 167–84.

14. Woodward, "The Army Medical Museum," 233–42; *Catalogue of the Surgical Section* (1866), 35; Brinton, *Personal Memoirs*, 189.

15. *Catalogue of the Surgical Section* (1866), 35; George Otis, *List of the Specimens in the Anatomical Section of the United States Army Medical Museum* (Washington, DC: Gibson Brothers, Printers, 1880), 150.

16. George Otis, *Check list of preparations and objects in the section of human anatomy of the United States Army Medical Museum. For use during the International exhibition of 1876, in connection with the representation of the Medical Department U.S. Army* (Washington, DC: Army Medical Museum, 1876), 100–106; Otis, *List of the Specimens*, 151.

17. Alexander Gardner, *Gardner's Photographic Sketchbook of the Civil War* (New York: Dover, 1959 [1866]), plate 94; Alan Trachtenberg, *Reading American Photographs: Images as History, Mathew Brady to Walker Evans* (New York: Hill & Wang, 1990), 110–11. See also Susan Mizruchi, "Becoming Multicultural," *American Literary History* 15, no. 1 (2003): 40–43. Neighbors of the Gettysburg battlefield also asked to be paid to bury the dead. Laderman, *Sacred Remains*, 103–9.

18. The battlefield dead sometimes left viewers confused about race. A *Harper's Weekly* correspondent stumbled among corpses at Antietam whose faces had turned "so black that no one would ever suspect that they had been white. All looked like negroes, and as they lay in piles where they had fallen, one upon another, they filled the by-standers with a sense of horror." "On Antietam Battle," *Harper's Weekly* 6 (1862): 655.

19. Louis Bagger, "The Army Medical Museum in Washington," *Appleton's Journal* 9 (March 1, 1873): 294–97; John Shaw Billings, *Description of Selected Specimens from the Army Medical Museum* (Chicago, 1892–93), 12–14; Woodward, "The Army Medical Museum," 233–42; *Catalogue of the Surgical Section* (1866), 35.

20. Woodward, "The Army Medical Museum," 242; Edward Anthony Spitzka, *A Study of the Brains of Six Eminent Scientists and Scholars Belonging to the American Anthropometric Society, Together with a Description of the Skull of Professor E. D. Cope* (Philadelphia: American Philosophical Society, 1907), 183.

21. 48th Congress, 1st session, Senate Executive Document, no. 12; "Army Medical Department," 1; the *National Police Gazette* reported in 1881 that in 1867

with President Johnson's permission, relatives removed Booth's remains from an unmarked grave in the Arsenal grounds to a family plot in a Baltimore cemetery. *National Police Gazette* 37 (February 5, 1881): 5. There were rumors in 1882 that the museum would display the body of President Garfield's assassin, Charles Guiteau. "Dr. Henry C. Yarrow denies the published report that the body of Guiteau, the assassin, has been dissected and the skeleton articulated for public exhibition. While declining to say where the body now is, he says it is ready to be turned over to the person whom the court before which the matter is now pending shall designate as entitled to it." "Notes from Washington," *New York Times*, December 11, 1882, 1. In June 1893 the old theater building collapsed under the weight of all the pension work, and a forty-foot section of the third floor gave way, killing twenty-three government clerks and injuring dozens of others. "A Crash in Ford's Theatre," *New York Times*, June 10, 1893, 1.

22. John Shaw Billings, "Medical Museums, with a Special Reference to the Army Medical Museum at Washington," The President's Address delivered before the Congress of American Physicians and Surgeons, September 20, 1888 (New Haven, CT: Tuttle, Morehouse & Taylor Printers, 1888), 370, 372, 377; Woodward, "The Army Medical Museum," 233–34, 240; Joanna R. Nicholls Kyle, "The Army Medical Library and Museum," *Godey's Magazine* 86 (April 1898): 409–17; Ernest Ingersoll, "The Making of a Museum," *Century Illustrated Magazine* 29 (January 1885): 359.

23. Billings, "Medical Museums," 372.

24. Otis, *List of the Specimens*, 4, 5.

25. Ibid., iv.

26. Ibid., 181, 51, 52, 105, 109, 123, 125, 131; McElderry to Barnes, October 25, 1873, NAA, box 5.

27. Otis, *List of the Specimens*, 150, 123.

28. Henry Adams, *The Education of Henry Adams: An Autobiography* (Boston: Houghton Mifflin, 1918), 351.

29. Benjamin Apthorp Gould, *Investigations in the Military and Anthropological Statistics of American Soldiers* (Cambridge, MA: Riverside Press, 1869), 5; J. H. Baxter, *Statistics, Medical and Anthropological of the Provost-Marshal-General's Bureau derived from Records of the Examination for Military Service in the Armies of the United States during the Late War of the Rebellion, over a million recruits, drafted men, substitutes, and enrolled men* (Washington, DC: Government Printing Office, 1875), ii–iv. Some scientific racists used the data on white men's short arms—information useful to jacket makers—to celebrate their distance from long-armed anthropoid apes. John Haller Jr., *Outcasts from Evolution: Scientific Attitudes of Racial Inferiority, 1859–1900* (Urbana: University of Illinois Press, 1971), 19–20. On insurance, see Viviana Zelizer, *Morals and Markets: The Development of Life Insurance in the United States* (New Brunswick, NJ: Transaction Books, 1983 [1979]), 67–89.

30. Baxter, *Statistics*, 84.

31. D. Wilson, *On the Supposed Prevalence*, 19; William Henry Jackson, *Descriptive Catalogue of Photographs of North American Indians* (Washington, DC: Government Printing Office, 1877), iii–iv.

32. Gould, *Investigations*, 308–11, 382.

33. Baxter, *Statistics*, 473. See also Peter T. Harstad, ed., "A Civil War Medical Examiner: The Report of Dr. Horace O. Crane," *Wisconsin Magazine of History* 48 (Spring 1965): 222–31. Winnebago and Chippewa peoples who witnessed the brutality of white reprisals against the Sioux may have calculated they were better off siding with Minnesota whites. On September 15, 1862, the *Davenport (IA) Gazette* reported, "Two Chippewa chiefs of Wisconsin, tender their services of Gov. Ramsey to fight the Sioux. It is well known that these two tribes have long been at war with each other." Seth J. Temple, *Camp McClellan during the Civil War: A Paper Read Before the Contemporary Club, Davenport, Iowa, October 22, 1927* (Davenport, IA: Contemporary Club, 1928), 21–22. A reporter for the *New York Times* suggested that the Winnebago murdered two Sioux men "probably in the hope of getting into favor with our people so they might be allowed to remain on their reservation." If this indeed was their motive, it didn't work. By the spring of 1863, the Winnebago were on their way to Fort Randall in the Dakota Territory. "The Rebel Sioux," *New York Times*, May 22, 1863, 1.

34. Gould, *Investigations*, 228, 237.

35. Alfred Muller, March 26, 1866, reel 2, NAA.

36. "The Rebel Sioux," 1; "Rebel Indians and Loyal Indians," *New York Times*, August 18, 1863, 4; Douglas Linder, "Famous American Trials: The Dakota Conflict Trials, 1862," http://www.law.umkc.edu/faculty/projects/ftrials/dakota/dakota.html (accessed February 21, 2009); Kenneth Carley, *The Sioux Uprising of 1862* (St. Paul: Minnesota Historical Society, 1976), 75; C. M. Oehler, *The Great Sioux Uprising* (New York: Oxford University Press, 1959), 222–23.

37. "The Rebel Sioux," 1.

38. "Rebel Indians and Loyal Indians," 4.

39. R. B. Hitz to J. K. Barnes, September 3, 1868, NAA, reel 1.

40. J. K. Barnes, "Circular no. 2," April 4, 1867, National Archives, RG 112, Records of the Office of the Surgeon General (Army) Central Office Issuance and Forms, 1818–1949, *Circulars and Letters of the Surgeon General's Office, 1861–1885, National Archives*, box 4. Assistant Surgeon Washington Matthews noted receiving Barnes's circular in the diary he kept at Fort Rice. Ray H. Mattison, ed., "The Diary of Surgeon Washington Matthews, Fort Rice, D.T.," *North Dakota History* (1954): 15.

41. Robert E. Bieder, "The Collecting of Bones for Anthropological Narratives," *American Indian Culture and Research Journal* 16, no. 2 (1992): 21–33; Robert E. Bieder, "The Representation of Indian Bodies in Nineteenth-Century Anthropology," in *Repatriation Reader: Who Owns American Indian Remains*, ed. Devon A. Mihesuah (Lincoln: University of Nebraska Press, 2000), 19–36; Robert E. Bieder, *A Brief Historical Survey of the Expropriation of American Indian Remains* (Boulder, CO: Native American Rights Fund, 1990); Fine-Dare, *Grave Injustice*, 30–33.

42. Sherry L. Smith, *The View from Officers' Row: Army Perceptions of Western Indians* (Tucson: University of Arizona Press, 1990), 2; Anton Paul Sohn, *A Saw, Pocket Instruments, and Two Ounces of Whiskey: Frontier Military Medicine in the Great Basin* (Spokane, WA: Arthur Clark Company, 1998).

43. Amateur engineers and architects sent recommendations for post housing for laundresses or for better methods for disposing of "offal, slops, and excreta." U.S. War Department, Surgeon General's Office, Washington, May 1, 1875, *A Report on the Hygiene of the United States Army with Descriptions of the Military Posts* (Washington, DC: Government Printing Office, 1875); Sohn, *A Saw*, 87; Maria Brace Kimball, *A Soldier-Doctor of Our Army: James P. Kimball* (Boston: Houghton Mifflin, 1917); William T. Corbusier, *Verde to San Carlos: Recollections of a famous Army Surgeon and his observant family on the Western frontier, 1869–1886* (Tucson, AZ: D.S. King, 1971); R. H. McKay, *Little Pills: An Army Story by R. H. McKay, formerly Acting Assistant Surgeon United States Army* (Pittsburg, KS: Pittsburg Headlight, 1918); Henry F. Hoyt, *A Frontier Doctor* (Boston: Houghton Mifflin, 1929); Rogers, *Selected Papers of John Shaw Billings*, 266.

44. George M. Kober, *Reminiscences of George Martin Kober, M.D. LL.D., emeritus dean and professor of hygiene of the School of medicine, and member of the Board of Regents, Georgetown University, Washington D.C.* (Washington, DC: Kober Foundation, 1930), 268. Kober sent the Paiute remains he unearthed to German ethnographer Rudolph Virchow, who included them in his *Crania Ethnica Americana*.

45. Kimball, *A Soldier-Doctor*, 106.

46. Otis, *List of the Specimens*, 146–55; George Williamson, *Observations on the Human Crania Contained in the Museum of the Army Medical Department, Fort Pitt, Chatham* (Dublin: McGlashan & Gill, 1857), 11.

47. Charles H. Crane, "Memorandum for the Information of Medical Officers," September 1, 1868, RG 112, box 4, National Archives; Lamb, "A History," 50–51, 89.

48. Billings, "Medical Museums," 368.

49. H. C. Yarrow, "Report on the Operation of a Special Party for Making Ethnological Researches in the Vicinity of Santa Barbara, Cal. with a short historical account of the region explored," in Engineer's Department, U.S. Army, *Report upon United States Geographical Surveys West of the One Hundredth Meriden, in charge of First Lieut. Geo. M. Wheeler* (Washington, DC: Government Printing Office, 1879), 35, 41; Lucien Carr, "Observations on the Crania from the Santa Barbara Islands, California," in Engineer's Department, ibid., 278; H. C. Yarrow, "Report of Collections Obtained from Cemeteries in the Vicinity of Santa Barbara," in Engineer's Department, ibid., 35; Mark Sibley Severence and Dr. H. C. Yarrow, "Notes upon the Human Crania and Skeletons Collected by the Expeditions of 1872–74," in Engineer's Department, ibid., 396.

50. J. N. Coleman, "George P. Hachenberg: An American Leonardo da Vinci," *Southwestern Historical Quarterly* 61 (1957–58): 464–73.

51. Otis, *List of the Specimens*, 121, 116, 118, 4, 110–13; George Hachenberg to J. K. Barnes, Fort Randall, D.T., January 18, 1869, NAA, reel 2, 481–506; Jerome A. Greene, *Fort Randall on the Missouri, 1856–1892* (Pierre: South Dakota Historical Society Press, 2005), 103.

52. Hachenberg to Surgeon General, Austin, Texas, October 20, 1879, reel 5, AMM, no. 2034, sect. IV, NAA; Otis, *List of the Specimens*, 110, 4.

53. Hachenberg to J. K. Barnes, Fort Randall, D.T., January 18, 1869; Otis, *List of the Specimens*, 111.

54. "Army Medical Department," 1; J. J. Woodward, "The Medical Staff of the United States Army, and its Scientific Work: An Address delivered to the International Medical Congress at Philadelphia, Wednesday evening September 6, 1876" (Philadelphia, 1876), 21, National Archives, box 3.

55. Otis, *List of the Specimens*, 145; William Adams to Billings, October 1887, reel 6, NAA; "Specimens Received," April 26, 1887, reel 6, NAA; Boas to J. S. Billings, April 23, 1887, reel 6, NAA; Boas in Bieder, *A Brief Historical Survey*, 45–46; Hayes to Otis, December 13, 1874, reel 5, NAA.

56. George Thomas Bettany, "Joseph Barnard Davis," in *The Dictionary of National Biography* (Oxford: Oxford University Press, 1968), 5:618–19; Joseph Barnard Davis, *Thesaurus Craniorum: Catalogue of the Skulls of the Various Races of Man in the Collection of Joseph Barnard Davis* (London: Printed for the Subscribers, 1867), 81, 151, 291; Joseph Barnard Davis, *Supplement to Thesaurus Craniorum: Catalogue of the Skulls of the Various Races of Man in the Collection of Joseph Barnard Davis, MD, FRS, FSA* (London: Printed for the Subscribers, 1875), 4, 15, 23, 33; "J.B.," "*Crania Britannica, Delineations and Descriptions of the Skulls of the Aboriginal and Early Inhabitants of the British Islands* by Joseph Barnard Davis; John Thurman," *Anthropological Review* 6 (January 1868): 52–55; MacDonald, *Human Remains*, 99–102.

57. Kiefer to Billings, June 30, October 10, December 5, 1888, and May 3, 1889, reel 6, NAA; *The George W. Kiefer Collection of Peruvian Antiquities, etc. etc.* (New York: Geo. A. Leavitt & Co., October 25, 1889). Billing's copy of the sale catalog is at the New York Public Library. Kiefer's story fit a pattern—a sort of curse on tomb robbers. For example, the *Galaxy* ran a story about a man who collected Pueblo dead. The collector accidentally shot himself three days after a local guide warned him "that those who disturbed the sleepers were sure to have bad luck or die by violence." "Homes of the American Aborigines," *The Galaxy: A Magazine of Entertaining Reading* 21 (April 1870): 521.

58. Paul Broca, *Instructions craniologiques et craniométriques* (Paris: Libraries Georges Masson, 1875); Gould, *The Mismeasure of Man*, 106.

59. Chapman, *Order Out of Chaos*, 319; John Shaw Billings, "On Composite Photography as Applied to Craniology" (Washington, DC: Government Printing Office, 1886); Otis, *List of the Specimens*, 104–5.

60. Billings, *On Composite Photography*, 106; "Composite Photography," *Manufacturer and Builder* 17 (July 1885): 147; Francis Galton, *Memories of My Life* (London: Metheun & Co., 1908), 244; Francis Galton, "Composite Portraits," *Nature* 18 (1878), 97–100; Francis Galton, "Photographic Composites," *Photographic News*, August 17, 1835, 243–45.

61. Alice Fletcher, "Composite Portraits of American Indians," *Science* 7, no. 170 (May 7, 1886): 408; Raphael Pumpelly, "Composite Portraits of Members of the National Academy of Sciences," *Science* 5, no. 118 (May 8, 1885): 378. Galton's biographer Karl Pearson was more skeptical about portraits of scientists, suggesting that the American composites of mathematicians, field geologists, and acade-

micians "lead us hardly further than the conclusion that all American scientists of those days were hairy, and that mathematicians while being least so had more frown." *The Life, Letters and Labours of Francis Galton*, 3 vols. (Cambridge: Cambridge University Press, 1924), 2:290n1.

62. Matthews, "Measuring the Cubic Capacity of Skulls," *Science* 5, no. 124 (June 19, 1885): 499–500; Billings, *On Composite Photography*, 105; Allan Sekula, "The Body and the Archive," *October* 39 (Winter 1986): 51.

63. Billings, *On Composite Photography*, 106–7, 111; Washington Matthews, "The Use of Rubber Bags in Gauging Cranial Capacity," *American Anthropologist* 11 (June 1898): 171–76.

64. Brinton, *Report upon the Collections*, 193.

65. Trouillot, *Silencing the Past*, 124–36.

66. David S. Snively to Surgeon General, October 4, 1879, reel 5, 2032, NAA; B. E. Fryer to Otis, February 12, 1869, reel 2, 509–26, NAA.

67. John Wesley Powell, "General Field Studies," *7th Annual Report of the Bureau of American Ethnology* (Washington, DC: Government Printing Office, 1891), xxix; "Explorations West of the 100th Meridian," *American Journal of Science and Arts* 5 (April 1873): 291; Otis, *List of the Specimens*, 51; "Biographies of People Honored in the Names of Amphibians and Reptiles of North America," http://ebeltz.net/herps/biogappx.html#XYZ (accessed December 10, 2008).

68. Henry Cercy Yarrow, *Introduction to the Study of Mortuary Customs among the North American Indians* (Washington, DC: Government Printing Office, 1880), 3, v; H. C. Yarrow, *A Further Contribution to the Study of the Mortuary Customs of the North American Indians* (Washington, DC: Government Printing Office, 1881); Powell, "General Field Studies," *7th Annual Report*, xxvii; "Indian Burial Customs," *Field and Stream* 15 (November 18, 1880): 302. See also David I. Bushnell Jr., *Burials of the Algonquian, Siouan, and Caddoan Tribes West of the Mississippi* (Washington, DC: Government Printing Office, 1927); and Roger C. Echo-Hawk, "Pawnee Mortuary Traditions," *American Indian Culture and Research Journal* 16, no. 2 (1992): 77–99.

69. James F. Russling, "National Cemeteries," *Harper's New Monthly Magazine*, August 1866, 310.

70. Yarrow, *Introduction*, 48–49; William Christie Macleod, "The Distribution and Process of Suttee in North America," *American Anthropologist* 33 (April–June 1931): 209–15. Traveler Richard Burton mentioned this burial in *The City of the Saints and across the Rocky Mountains to California* (New York: Harper & Brothers, 1862), 475. Yarrow's letter of February 12, 1873, gave the excavation more life. Lt. Mott and two soldiers helped him dig up the remains. Yarrow to Otis, reel 3, 966, NAA.

71. Will Bagley, *Blood of the Prophets: Brigham Young and the Massacre at Mountain Meadows* (Normal: University of Oklahoma Press, 2004), 32.

72. George McKenzie, "Cause and Origin of the Walker War," in *History of Indian Depredations in Utah*, ed. Peter Gottfredson (White Fish, MT: Kessinger, 2006 [1919]), 43–47; "Arrivals from the Salt Lake City—The Walker War—Indian Fighting and Starving," *Ripley (OH) Bee*, March 18, 1854; and Gustive Larson,

"Wakara's Half Century," *Western Humanities Review* 6, no. 3 (Summer 1952): 235–59. Young is quoted in Solomon Nunes Carvalho, *Incidents of Travel and Adventure in the Far West* (New York: Derby & Jackson, 1860), 52; and Blackhawk, *Violence over the Land*, 236–44. On the context for this war, see Jared Farmer, *On Zion's Mount: Mormons, Indians, and the American Landscape* (Cambridge, MA: Harvard University Press, 2009), 19–53.

73. Carvalho, *Incidents*, 189–90.

74. Gottfredson, *History of Indian Depredations*, 84.

75. Yarrow to Charles H. Crane, February 12, 1873, box 6, NAA; Yarrow to Otis, February 20, 1873, box 6, NAA; Otis, *List of the Specimens*, 51.

EPILOGUE

1. "Dissecting the Brains of 100 Famous Men for Science," *New York Times*, September 29, 1912; Spitzka, *A Study of the Brains of Six Eminent Scientists and Scholars*, 175–76. Spitzka's study collected reports on brains of composers Beethoven, Schumann, and Donizetti; naturalists George Cuvier and Morton's friend Louis Agassiz; ethnologist John Wesley Powell; phrenologist Franz Joseph Gall; brain scientist Paul Broca; statesmen Daniel Webster, Léon Gambetta, and Napoleon III; writers William Makepeace Thackeray, the astoundingly big-brained Ivan Turgenev, and Walt Whitman (whose brain was dropped and broken by a careless attendant); along with several mathematicians, physiologists, and physicians (177–206). The widow of physiognomist Joseph Simms made sure that Spitzka got his brain. "Body of Lecturer Given to Science," *New York Times*, April 13, 1920.

2. See Burrell, *Postcards from the Brain Museum*; and Paterniti, *Driving Mr. Albert*.

3. Anthropologist Orin Starn traces this history in *Ishi's Brain*. Many of us first read about Ishi in Theodora Kroeber, *Ishi in Two Worlds: A Biography of the Last Wild Indian in North America* (Berkeley: University of California Press, 1961). Her husband, anthropologist Kroeber, one of the faculty members who was closest to Ishi, was in Germany when Ishi died and was unable to intervene before the brain was taken. Aleš Hrdlička describes his interest in brains in *An Eskimo Brain* (New York: Knickerbocker Press, 1901). For a moving account of the impact of Hrdlička's collecting, see Harper, *Give Me My Father's Body*.

4. M. F. Ashley Montagu, "Aleš Hrdlička, 1869–1943," *American Anthropologist* 46 (1944): 113–17; Stephen Loring and Miroslav Prokopec, "A Most Peculiar Man: The Life and Times of Aleš Hrdlička," in *Reckoning with the Dead: The Larsen Bay Repatriation and the Smithsonian Institution*, ed. Tamara L. Bray and Thomas W. Killion (Washington, DC: Smithsonian Institution Press, 1994), 26–53.

5. Aleš Hrdlička to Edward J. Nolan, May 2, 1911, Morton Papers, APS. See Aleš Hrdlička, "Catalogue of Human Crania in the United States National Museum Collections," *Proceedings of the United States Museum* 63 (1923): 1–51; Hrdlička, *Physical Anthropology*, 41.

6. Aleš Hrdlička, *Skeletal Remains Suggesting or Attributed to Early Man in North America* (Washington, DC: Government Printing Office, 1907), 11.

7. Hrdlička, *Physical Anthropology*, 8; Aleš Hrdlička, "Arrangement and Preservation of Large Collections of Human Bones for Purposes of Investigation," *American Naturalist* 39 (January 1900): 9–15.

8. Hrdlička, *Physical Anthropology*, 15, 17; Frank Spencer, ed., *A History of American Physical Anthropology, 1930–1980* (New York: Academic Press, 1982), 6; C. Loring Brace, "The Roots of the Race Concept in American Physical Anthropology," in *A History*, ed. Spencer, 11–29.

9. Loring and Prokopec, "A Most Peculiar Man," 31–32, 36; Aleš Hrdlička, *Alaska Diary, 1926–1931* (Lancaster, PA: Jacques Cattell Press, 1943), 16, 115, 127, 129–30, 223.

10. Brown, *Reaper's Garden.*

11. Loring and Prokopec, "A Most Peculiar Man," 32; Hrdlička, *Alaska Diary*, 10.

12. Anna C. Roosevelt, John Douglas, and Linda Brown, "The Migrations and Adaptations of the First Americans: Clovis and Pre-Clovis Viewed from South America," in *The First Americans: The Pleistocene Colonization of the New World*, ed. Nina G. Jablonski (San Francisco: California Academy of Sciences, 2002), 159–236.

13. R. W. Shufeldt, *Chapters on the Natural History of the United States* (Glacier, MT: Kessinger Publications, [1897]), 28; R. W. Shufeldt, "Life History of American Naturalist," *Medical Life* 31, no. 3 (March 1924): 115; *Medical Life* 31, no. 4 (April 1924): 143–47; and *Medical Life* 31, no. 5 (May 1924): 195; Kalman Lambrecht, "In Memoriam: Robert Wilson Shufeldt, 1850–1934," *The Auk: A Quarterly Journal of Ornithology* 52 (October 1935): 359–61.

14. Shufeldt, "Life History," (May 1924): 192.

15. R. W. Shufeldt, "Personal Adventures of a Human Skull Collector," *Medical Council* 15, no. 4 (April 1910): 123–27; R. W. Shufeldt, "A Navajo Skull," *Journal of Anatomy and Physiology* 20 (April 1886): 425–29; R. W. Shufeldt, "The Skull of a Navaho Child," *Journal of Anatomy and Physiology* 21 (October 1886): 66–71; R. W. Shufeldt, "Contribution to the Comparative Craniology of the North American Indians: the Skull in the Apaches," *Journal of Anatomy and Physiology* 21 (1887): 525–35.

16. Shufeldt, "Life History," (March 1924): 105–6, 109–13; (May 1924): 193, 200.

17. R. W. Shufeldt, *America's Greatest Problem: The Negro* (Philadelphia: F.A. Davis Company, 1915), 184; "America's Greatest Problem: The Negro," *New York Times*, July 18, 1915.

18. Andrew Gulliford, "Bones of Contention: The Repatriation of Native American Human Remains," *Public Historian* 18, no. 4 (1996): 119–43. The controversy over "Kennewick Man," the 9,000-year-old skeleton found in Washington State in 1996, has made some of these issues particularly visible. See Jeff Benedict, *No Bone Unturned: The Adventures of a Top Smithsonian Forensic Scientist and the Legal Battle for America's Oldest Skeletons* (New York: HarperCollins, 2003); and Thomas, *Skull Wars*, xvii–xxxix, 231–35.

19. National NAGPRA, "Laws, Regulations, and Guidance," http://www.nps.gov/history/nagpra/mandates/index.htm (accessed March 5, 2009); Kathleen

Fine-Dare, "Histories of the Repatriation Movement," in *Opening Archaeology: Repatriation's Impact on Contemporary Research and Practice*, ed. Thomas W. Killion (Santa Fe: School for Advanced Research Press, 2007), 29–56.

20. Preston, "Skeletons in Our Museums' Closets," 68.

21. Kara Swisher, "Skeletons in the Closet," *Washington Post*, October 3, 1989, D5; Larry J. Zimmerman, "Archaeology, Reburial, and a Discipline's Self-Delusion," *American Indian Culture and Research Journal* 16 (1992): 37–56; Larry J. Zimmerman, "Remythologizing the Relationship between Indians and Archaeologists," in *Native Americans and Archaeologists: Stepping Stones to Common Ground*, ed. Nina Swidler, Kurt E. Dongoske, Roger Anyon, and Alan S. Downer (Walnut Creek, CA: AltaMira Press, 1997); James Riding In, "Six Pawnee Crania: Historical and Contemporary Issues Associated with the Massacre and Decapitation of Pawnee Indians," *American Indian Culture and Research Journal* 16, no. 2 (1992): 101–19; Thomas, *Skull Wars*, 209–24; Vine Deloria, "Secularism, Civil Religion, and the Religious Freedom of American Indians," *American Indian Culture and Research Journal* 16, no. 2 (1992); Gary White Deer, "Return of the Sacred: Spirituality and the Scientific Imperative," in *Native Americans and Archaeologists*, ed. Swidler, Dongoske, Anyon, and Downer, 37–43.

Index

Page numbers in italics refer to illustrations. SGM refers to Samuel George Morton.